# The Latin American Studies Book Series

*The Latin American Studies Book Series* promotes quality scientific research focusing on Latin American countries. The series accepts disciplinary and interdisciplinary titles related to geographical, environmental, cultural, economic, political, urban and health-related research dedicated to Latin America. The series publishes comprehensive monographs, edited volumes and textbooks refereed by a region or country expert specialized in Latin American studies.

The series aims to raise the profile of Latin American studies, showcasing important works developed focusing on the region. It is aimed at researchers, students, and everyone interested in Latin American topics.

Submit a proposal: Proposals for the series will be considered by the Series Advisory Board. A book proposal form can be obtained from the Publisher, Andrea Brody-Barre (andrea.brody-barre@springer.com).

Liliana Lizárraga-Mendiola ·
Carlos Alfredo Bigurra-Alzati ·
Gabriela A. Vázquez-Rodríguez
Editors

# Sustainable Spaces in Arid and Semiarid Zones of Mexico

Springer

*Editors*
Liliana Lizárraga-Mendiola
Autonomous University of Hidalgo State
Mineral de la Reforma, Hidalgo, Mexico

Carlos Alfredo Bigurra-Alzati
Autonomous University of Hidalgo State
Mineral de la Reforma, Hidalgo, Mexico

Gabriela A. Vázquez-Rodríguez
Autonomous University of Hidalgo State
Mineral de la Reforma, Hidalgo, Mexico

ISSN 2366-3421 ISSN 2366-343X (electronic)
The Latin American Studies Book Series
ISBN 978-3-031-53417-1 ISBN 978-3-031-53418-8 (eBook)
https://doi.org/10.1007/978-3-031-53418-8



This Springer imprint is published by the registered company Springer Nature Switzerland AG
The registered company address is: Gewerbestrasse 11, 6330 Cham, Switzerland

# Contents


1 **Toward Sustainable Communities in Arid and Semi-Arid Zones of the Global South** ........ 1
Liliana Lizárraga-Mendiola, Carlos Alfredo Bigurra-Alzati, and Gabriela A. Vázquez-Rodríguez

2 **Sustainable Cities and Urban Morphology in Arid and Semi-Arid Climates of Mexico** ........ 17
Alma Delia Juárez-Sedano and Miguel Ángel González-Loza

3 **Construction of the Urban Sustainability Index for Cities in Mexico and Its Application in the Municipalities of Pachuca and Mineral De La Reforma, 2020** ........ 33
Patricia Catalina Medina-Pérez, Jorge Villanueva-Solís, and Sergio Gabriel Ceballos-Pérez

4 **Unearthing Disappeared Cultural Landscapes: The Case of the Old Tram Route of the Pachuca-Real Del Monte Mining District, Hidalgo, Mexico** ........ 51
Carlos Alfredo Bigurra-Alzati, María Elena Sánchez-Roldán, Valeria Volpi-León, and Julio César García-Martínez

5 **Blue and Green Infrastructure: History and Experiences in Mexico and the Arid and Semi-Arid Global South** ........ 69
Gabriela A. Vázquez-Rodríguez, Marcelino Antonio Zúñiga-Estrada, and Joyce E. Ortiz-Hernández

**6 Traditional Construction Systems and Their Sustainable Opportunities: Adobe and Rammed Earth** . . . . . . . . . . . . . . . . . . . . . . . . 91
Luis Daimir López-León, Alejandro González-de-la-Fuente, and Liliana Lizárraga-Mendiola

**7 Sustainable Technologies Applied to Asphalt Paving by Reusing Polymeric Waste** . . . . . . . . . . . . . . . . . . . . . . . . . . . . . . . . . . . . . . . . . . . . . . 111
Iván Erick Castañeda-Robles, Francisco Javier Olguín-Coca, and Citlali Gaona-Tiburcio

# Contributors

**Carlos Alfredo Bigurra-Alzati** Engineering and Architecture Academic Area, Autonomous University of Hidalgo State, Mineral de la Reforma, Hidalgo, Mexico

**Iván Erick Castañeda-Robles** Engineering and Architecture Academic Area, Autonomous University of Hidalgo State, Mineral de la Reforma, Hidalgo, Mexico

**Sergio Gabriel Ceballos-Pérez** National Council for Science and Technology (CONACYT), Ciudad de México, Mexico

**Citlali Gaona-Tiburcio** Faculty of Mechanical and Electrical Engineering, Autonomous University of Nuevo León, San Nicolás de los Garza, Nuevo León, Mexico

**Julio César García-Martínez** Civil Engineering, National Technological Institute of Mexico, Pachuca de Soto, Hidalgo, Mexico

**Alejandro González-de-la-Fuente** Engineering and Architecture Academic Area, Autonomous University of Hidalgo State, Mineral de La Reforma, Hidalgo, Mexico

**Miguel Ángel González-Loza** Higher Public Normal School of the State of Hidalgo, Pachuca de Soto, Hidalgo, Mexico

**Alma Delia Juárez-Sedano** Engineering and Architecture Academic Area, Autonomous University of Hidalgo State, Mineral de la Reforma, Hidalgo, Mexico

**Liliana Lizárraga-Mendiola** Engineering and Architecture Academic Area, Autonomous University of Hidalgo State, Mineral de La Reforma, Hidalgo, Mexico

**Luis Daimir López-León** Engineering and Architecture Academic Area, Autonomous University of Hidalgo State, Mineral de La Reforma, Hidalgo, Mexico

**Patricia Catalina Medina-Pérez** Sustainable Urban Development, The College of the State of Hidalgo, Hidalgo, Mexico

**Francisco Javier Olguín-Coca** Engineering and Architecture Academic Area, Autonomous University of Hidalgo State, Mineral de La Reforma, Hidalgo, Mexico

**Joyce E. Ortiz-Hernández** Iztépetl Ciencias de la Tierra y Afines S.C., Pachuca, Hidalgo, Mexico

**María Elena Sánchez-Roldán** Engineering and Architecture Academic Area, Autonomous University of Hidalgo State, Hidalgo, Mexico

**Jorge Villanueva-Solís** School of Architecture, Autonomous University of Coahuila, Torreón, Coahuila, Mexico

**Valeria Volpi-León** Engineering and Architecture Academic Area, Autonomous University of Hidalgo State, Mineral de la Reforma, Hidalgo, Mexico

**Gabriela A. Vázquez-Rodríguez** Chemistry Academic Area, Autonomous University of Hidalgo State, Mineral de la Reforma, Hidalgo, Mexico

**Marcelino Antonio Zúñiga-Estrada** Earth Sciences and Materials Academic Area, Autonomous University of Hidalgo State, Mineral de la Reforma, Hidalgo, Mexico

# Chapter 1
# Toward Sustainable Communities in Arid and Semi-Arid Zones of the Global South

**Liliana Lizárraga-Mendiola, Carlos Alfredo Bigurra-Alzati, and Gabriela A. Vázquez-Rodríguez**

**Abstract** Human beings rely on environment modification for the proper satisfaction of needs such as shelter building, food and tool production, water supply, sanitization, and recreation. Environmental changes brought about by human societies have reached a breakthrough point known as Anthropocene, one of the characteristics of which is urbanization. Given that from 2007 most world population is concentrated in urban settlements, decisions taken by authorities, organizations, and citizens in general concerning the urban spaces can constitute a driving force toward sustainability. Construction of sustainable spaces is more urgent in arid and semi-arid zones of the world, which cover around 41% of the world's land surface, are inhabited by 38% of the global population, and are considered among the most vulnerable regions to climate change. This book aims to highlight opportunities and acquired experiences in applying urban planning strategies, eco-building practices, and low-impact and passive technologies, to the sustainable management of urban spaces in arid and semi-arid regions of Mexico.

**Keywords** Urbanization · Unsustainability · Sustainable development goals · Natural corridors · Blue and green infrastructure

---

L. Lizárraga-Mendiola (✉) · C. A. Bigurra-Alzati
Engineering and Architecture Academic Area, Autonomous University of Hidalgo State, Carr. Pachuca-Tulancingo Km 4.5, Col. Carboneras. Mineral de La Reforma, C.P. 42184 Hidalgo, Mexico
e-mail: mendiola@uaeh.edu.mx

C. A. Bigurra-Alzati
e-mail: carlos_bigurra@uaeh.edu.mx

G. A. Vázquez-Rodríguez
Chemistry Academic Area, Autonomous University of Hidalgo State, Carr. Pachuca-Tulancingo Km 4.5, Col. Carboneras. Mineral de La Reforma, C.P. 42184 Hidalgo, Mexico
e-mail: gvazquez@uaeh.edu.mx

 
L. Lizárraga-Mendiola et al. (eds.), *Sustainable Spaces in Arid and Semiarid Zones of Mexico*, The Latin American Studies Book Series,
https://doi.org/10.1007/978-3-031-53418-8_1

## 1.1 Introduction

Since its origin, humankind has striven to manipulate the environment to satisfy its needs in livable spaces. Our lives occur in or on civil infrastructure such as houses, offices, schools, streets, or malls, among many others. Cement consumption is considered an indicator of the economic growth of a country, and the construction industry is one of the primary sources of development for any nation. In 2019 the world construction market was estimated at 11.4 trillion USD, which had grown at an annual rate of 3.71% during the five previous years (Market Prospects 2021). In addition, 17% of the world's total employed population works in this industry (ConstructionPlacements 2021).

Indeed, our existence depends on all kinds of built spaces, which are increasingly concentrated in cities. These are home to more than half of the world's population, which is expected to increase to nearly two-thirds by 2050 (UN 2019). Cities are at the center of global economic forces, linking places near and far through extensive exchange networks. While this increased urbanization has meant more socioeconomic opportunities and social well-being for some, it also increases pressure on surrounding ecosystems. For instance, cities contribute more than 75% of the global gross domestic product but account for over 70% of the world's $CO_2$ emissions, with transport and buildings being the major contributors (Elmqvist et al. 2019; WMO 2022). Since urbanization is occurring worldwide, it is considered a driver of the Anthropocene, that is, the epoch we live in characterized by deep earth biophysical changes concurring with the accelerating rise in human activities (Elmqvist et al. 2021). Cities, with their enormous ecological footprint, embody the quintessential Anthropocene landscape. Growing urban resource demand results from population or infrastructure size, consumption patterns, lifestyles, urban-centered policies, and management decisions (Elmqvist et al. 2021). These factors often also lead to traffic congestion, criminality, or systemic inequities, among other indicators of low life quality.

By persisting for thousands of years, cities have proven to be resilient socio-environmental systems. However, with the scale change implicit in the Anthropocene concept, urban agglomerations are increasingly vulnerable to the climate change impacts such as extreme rainfall events, droughts, or heat stress. About 90% of the urban areas are situated on coastlines, and low-lying agglomerations are likely to face more frequent and more extensive coastal flooding due to sea level rise, storm surges, and subsidence (Elmqvist et al. 2021; WMO 2022).

Further, cities are facing more frequent and intense heat waves compared to conditions prevailing during the 1950s, and it is estimated that over 1.6 billion people living in approximately 970 cities will be exposed in a regular way to three-month average temperatures of at least 35 °C by the 2050s (WMO 2022). Highly-concentrated populations, services provision, and interlinked infrastructures make cities more prone to human and economic losses than rural areas when disasters occur. Urban infrastructure is so concentrated and interconnected nowadays that there is a continuous risk of overall cascading failure of essential services such as water or energy supply, sewage collection, or transportation (Elmqvist et al. 2021).

Most urban growth occurs in the Global South, which covers 60% of the world's terrestrial surface and houses around 80% of the world's population (Shackleton et al. 2021). The catch-all term Global South has been increasingly used by scholars of various fields, from international relations to social and natural sciences (Shackleton et al. 2021). It is the most recent addition to a family of terms, including third world, peripheric, developing, or undeveloped countries, used to identify and cluster the low-income and politically-marginalized regions of the world (Clarke 2018). Global South is key in addressing urban unsustainability at the global level because, nowadays, the predominant urbanization form is the enlargement of megalopolis such as Mexico City, São Paulo, or Johannesburg-Soweto (Castells 2002). Besides, in the Global South cities, this urban growth occurs in an uncertain context marked by climate change, informal economies, housing crisis, deterioration of public hygiene, and extractivist offensives from the Global North economies, among many other factors.

Next, we will elaborate further on the unsustainability of current cities while focusing on the construction industry. This context will be followed by a brief state of the art of proposals for building sustainable and resilient communities using the Sustainable Development Goals (SDGs) as a roadmap and a presentation of the particularities of the arid and semi-arid zones of the Global South, which require tailor-made efforts aimed at the same purpose. Finally, we detail the scope of this book.

### *1.1.1 Urbanization and Unsustainability*

The dissemination and growth of the population on the planet have wreaked havoc on biodiversity, which includes the extinction of plant and animal species, in addition to the deterioration and decrease of environments or natural sites that are increasingly scarce (McGrath 2019). The first human settlements began about 14,000 years ago when humans developed agriculture. At that time, society's environmental impact was minimal, so there is no record of atmospheric emissions (Foundation BNP Paribas 2021). Later, with the invention of the wheel 3500 years ago, humans could transport heavy loads to other places, which impacted the environment with 23% of global $CO_2$ emissions. As technological advances evolved, there was an increase in atmospheric emissions. For example, in 1750, with the beginning of the Industrial Revolution, the relationship between anthropogenic activities and the generation of pollutants and atmospheric emissions strengthened (Foundation BNP Paribas 2021). For this reason, several authors, including Paul Crutzen, one of the leading proponents of the Anthropocene epoch, consider the Industrial Revolution its starting point (Crutzen 2002). For other authors, it could have begun around 1945, when $CO_2$ emissions, as well as an extensive list of other indicators (e.g., world population, fertilizers consumption, or ocean acidification), started to increase exponentially (Zalasiewicz et al. 2015). The Great Acceleration term was proposed to describe the departure point of this escalation in human activities that followed distinctive patterns of population growth and resource use based on fossil fuels (Steffen et al. 2004).

One of the results of these patterns is that, in 2020, the mass of human-made matter exceeded that of living biomass and reached about 1154 Tt (Elhacham et al. 2020). Of this amount, at least 94% corresponds to materials directly associated with the construction industry, such as concrete, bricks, and asphalt. Its scale is a significant factor explaining why this industry is unsustainable. Since annual concrete production is estimated at 30 billion tons, it exceeds the per capita production of any other material (Monteiro et al. 2017). Manufacturing this massive amount of concrete requires a proportional entry of aggregates (sand, gravel, and crushed rock), whose extraction has increased drastically since 1950 (Torres et al. 2021). A conservative estimate of the sand extraction is 40 Gt annually (Peduzzi 2014), making it the second most used natural material after water and the most mined material globally (Bisht 2021). Furthermore, buildings and construction account for 36% of global energy use and 39% of $CO_2$ emissions annually (Budds 2019). Construction creates around a third of the global waste (Miller 2021) and contributes to 40% and 23% of the drinking water pollution and the deterioration of urban air quality, respectively (Dixon 2010).

At the same time, while these atmospheric emissions contribute to global warming, ecosystem changes occurring in the Anthropocene can impact the built environment in multiple ways. Therefore, it is vital to adapt construction techniques to different segments of society and prevailing climatic conditions to sustainably meet the needs of a constantly growing population (Andrić et al. 2019). This is particularly urgent in many parts of the Global South, where urban planning and governance are significantly limited by inadequate investment, lack of properly-skilled personnel, and the predominance of reactive or retrospective planning over prospective approaches (Shackleton et al. 2021). In addition, not only the world population growth moves us away from the objectives of international treaties such as the Paris Agreement, which aims to achieve zero net emissions of greenhouse gases by 2030, but also the current consumption levels (IPCC 2018). Consequently, the built environment becomes even more vulnerable to overexploitation of natural resources, rising sea levels, droughts, floods, and other weather events (Hürlimann et al. 2022).

## 1.2 Sustainability and the Built Environment

Due to the dominance of urban growth, it is becoming increasingly clear that cities have a crucial role in achieving global sustainability goals because of their potential to address and impact issues such as climate change, biodiversity loss, and water conservation through innovative planning, inclusive decision-making, and multi-level governance (Paterson et al. 2015; WMO 2022).

Urban development today should not only focus on building more infrastructure but also on adopting measures that preserve natural resources and, at the same time, guarantee the well-being of the city's inhabitants. For example, in cities where planning is inadequate, the saturation of spaces and the demand for services to meet the population's needs are exceeded. That is why urban planning focused on social and environmental aspects can help resolve community conflicts and use its available resources and infrastructure more efficiently (Campbell 1996). Other authors propose to restore existing urban centers through the reconfiguration of sprawling suburbs where natural environments are preserved and the identification of the existing urban conflicts to detect the gaps that allow for solving the urban inequity of the less-favored societal groups (Godschalk 2004). Another study focused on determining the environmental and social equity in low-income urban areas, especially for social minority groups (Zhang et al. 2022). Therefore, sustainable strategies inside urban planning could be adopted by authorities, scholars, constructors, and society to mitigate a population's environmental and socioeconomic impacts with limited options to guarantee their well-being.

For more than 30 years, several studies have mentioned the displacement of urban limits due to population growth. This urban development transformed natural landscapes into sprawling neighborhoods at an unprecedented rate. Buck et al. (2021) observe an inverse relationship between the segregation of society and sustainability. The strategies adopted by governments, industry, and researchers have included from health and education issues to more complex social cohesion and equity issues. A fundamental principle of urban sustainability is equal access and opportunities for everyone (Opp 2017). If any inhabitant lacks access to public services and cannot meet their basic needs, then urban sustainability is non-existent (Sampson 2017). Free and accessible common spaces can positively affect the health of socioeconomically disadvantaged groups since they are an alternative resource when other health services are lacking (Moran et al. 2021).

Peri-urban zones often include natural areas and ecological corridors that deserve special attention because they can become negatively impacted, fragmented, or inhospitable for many species around densely urbanized sites due to the decrease in vegetation cover (Pannunzio-Ribeiro et al. 2022). That is why the conservation of these ecological corridors and their historical legacy is within the scope of urban sustainability. These sites have been proven to be landscapes of memory and have value in preserving nature and culture (Woudstra and Fieldhouse 2000). United Nations Educational, Scientific and Cultural Organization (UNESCO) defines cultural heritage as "works of man or the combined works of nature and man, including archaeological sites, which are of outstanding universal value from the historical, aesthetic, ethnological or anthropological point of view" (WHC 2017). When these areas are conserved and valued, they can provide ecosystem services and accessible recreation areas for the population (Pannunzio-Ribeiro et al. 2022). Therefore, maintaining adequate connectivity in these corridors considered anthropic landscapes favors the care of local biodiversity, ensuring the persistence of local species through the dispersion of the various organisms that coexist there (Beier and Noss 1998). Moreover, it has been proven that caring for the connectivity of peri-urban ecological corridors can be compatible with organized urban growth and economic development (Hernández et al. 2015).

The restriction of green areas due to the built environment's growth reduces their capacity to provide ecosystem services (Cueva et al. 2022). These services are essential to guarantee the conservation of natural resources since they provide spaces for the permanence of plant and animal species. They also offer spaces for the infiltration of urban runoff, reduce the risk of flooding, enable other sources of collection and supply of water to the population, and increase the real state value, among other recognized sustainable advantages (Jim and Chen 2006). Consequently, blue and green infrastructure (BGI) is considered an option in sustainable urban planning, which consists of strategically-planned networks of semi-natural or restored spaces that guarantee the provision of the aforementioned ecosystem services (Macháč et al. 2022).

Although studies on BGI are vast, their approach is heterogeneous and addresses different needs separately. Bellezoni et al. (2021) mention that in some cities of the Global South, the priority has been to address the link between BGI and the water-energy-food nexus, while other urban sites focus on the cooling effect of the green space, flood reduction, and water capture for supply. For example, BGI has been explored in semi-arid areas of Latin America to mitigate high temperatures inside homes and capture and store rainwater for domestic purposes (Melo-dos-Santos et al. 2019). In Chinese cities, gray infrastructure is being optimized with adaptations of BGI to control runoff (Liu et al. 2021). This approach has allowed to mitigate the impact of floods and improve groundwater recharge at once.

Recent efforts to improve sustainability in cities focus on increasing the quantity and quality of BGI. These measures enhance air quality and outdoor recreation spaces and drive economic improvements (De Sousa 2003). However, despite the multiple benefits mentioned above, it is vital to monitor and try to avoid the displacement of the low-income population elsewhere, which constitutes an environmental injustice example that accompanies an already chronic social inequality (Easton et al. 2020). This occurs due to the increase in the surplus value of these revitalized spaces, which may be inaccessible for many, causing the phenomenon of environmental or green gentrification. There are few current attempts to comprehensively analyze the connections between environmental quality, social segregation, and sustainability across entire urban areas. For instance, Moran et al. (2021) identified that the presence of green spaces in Latin American cities could reduce educational inequalities and the mortality of its inhabitants. However, the practical implementation of these findings and the actual reduction of social inequities are still lacking in this region.

As stated above, the construction sector demands large amounts of natural resources. This extraction entails challenges such as protecting stone resources from overexploitation since they have significant ecological value by providing essential ecosystem services. For example, they protect against floods, promote food production, and store and filter groundwater (Torres et al. 2021). Among the alternatives to reduce this overexploitation is substituting conventional materials (concrete and steel) with local natural resources that require less fuel consumption during transportation and production. Eco-building with adobe and rammed earth is an option to minimize materials costs for transportation and fuel consumption. According to a cradle-to-grave analysis, these construction processes can mitigate up to 50% of the environmental emissions produced by traditional construction processes (Fernandes et al. 2019). Adobe and rammed earth techniques are more sustainable compared to concrete because their embodied energy consumption (5–10 kWh/m$^3$) is lower compared to cement (2640 kWh/m$^3$) (Charai et al. 2022). Their adoption promotes the employment of passive technologies in the design of modern buildings that achieve greater energy efficiency and, at the same time, provide comfort to the population (Charai et al. 2022).

Another eco-building alternative to reduce the consumption of natural aggregates is diminishing their use in communication routes or highways. Although the construction and maintenance of roads are critical factors in any region's economic development, the stone aggregates consumed to build this type of infrastructure exert considerable pressure on the environment (Bamigboye et al. 2021). Given this, various studies have used residues such as polyethylene terephthalate in plastic bottles to reverse the high consumption of gravel and sand with positive results. These residues have been used as light aggregates, geomembranes, and asphalt binders (Bamigboye et al. 2021). In addition to giving favorable results in the construction sector, its reuse represents positive environmental impacts since only 9% of this waste is recycled (Hassan et al. 2004). Another study showed that reusing tires on asphalt pavement reduced energy use during materials' production and construction phases by 36% and 45%, respectively, compared to virgin materials (Farina et al. 2016). Also, it was shown that the costs decreased due to the shorter distances required for transporting the materials to the construction site.

### *1.2.1 The Sustainable Development Goals as a Roadmap to Sustainability*

The publication of the Brundtland report in 1987 marked the rise of the sustainable development paradigm as an institutional principle and policy goal and a fundamental shift in thinking in development, environment, and governance (Sneddon et al. 2006). Notably, in the sustainable development concept was first introduced the tension between the needs of future generations and the current unmet needs of most of the world's population. However, while the importance of the concept was accepted and even saluted, its operation on the ground was placed in doubt by many.

The ambitious, and often labeled as too vague, objectives of sustainable development were concretized in a set of targets known as the United Nations Millennium Development Goals. They were updated and enlarged in 2015 as the 17 SDGs, having 2030 as a time horizon to eradicate poverty and hunger, improve life quality, and safeguard the environment. Each SDG has 169 associated targets and 230 indicators, whose advance is monitored nationally and globally in a way that, as a whole, SDGs constitute not only an aspirational but a practical roadmap to sustainability. Due to the intertwinement between SDGs, strategies aiming to fulfill them must be developed at the national level by considering interdependence and eventual trade-offs, as well as the contributions of science, technology, and innovation and the insights from social sciences.

## 1.3 Particularities of the Global South's Arid and Semi-Arid Cities

Arid and semi-arid areas can be broadly categorized by annual precipitation rates lower than 500 mm and potential evapotranspiration higher than precipitation (D'Odorico et al. 2019). These zones are prone to extreme events such as droughts and land degradation due to water scarcity. This degradation may originate from natural or anthropogenic causes and lead, in some cases, to desertification. The world's arid and semi-arid zones lie mainly (72%) in the Global South and are inhabited by more than 2000 million people (Huber-Sannwald et al. 2012). Furthermore, arid and semi-arid areas are expected to enlarge due to climate change, and they are considered the most vulnerable to its effects (Spinoni et al. 2021). For instance, in Mexico, around 50% of the land, housing 30% of its population, is characterized by arid and semi-arid climates (Huber-Sannwald et al. 2012). In a climate change scenario, it has been pointed out that precipitation will gradually decrease almost everywhere in the country compared to the historical data recorded for 1961–1990, with more significant consequences in regions already experiencing water scarcity. Likewise, an increase in the frequency and intensity of extreme weather events such as cyclones, torrential rains, and droughts has been predicted for these areas (Ortiz-Hernández 2016). The climate change effects in the arid and semi-arid zones of Mexico concur with many other environmental, social, and economical problems that are just as or more relevant, as it occurs generally in the Global South.

The Global South idea contests the belief of the homogenization of world cultures and societies carried out by international trade and, on the contrary, designates the territories and spaces negatively impacted by globalization. It is not an idea centrally focused on development or cultural differences within the world nations, but rather on the underlying geopolitical relations of power (Dados and Connell 2012). At the same time, it is a vindicatory concept that draws attention to the common struggles experienced by diverse societies and groups under contemporary capitalism while recognizing their agency to establish their own initiatives and policy responses (Clarke 2018).

Likely, the leading Global South challenge is epistemological because, even internally, the region is not seen as a source of the solutions to its own problems. Indeed, the local solutions are conceived according to knowledge produced in the Global North. In the field of urban studies, Nagendra et al. (2018) have stressed that, although the Global South urban issues are different from those of the Global North, our current urban knowledge is shaped and biased by research made in the Global North. This must change, given the importance of the local contexts and the participation of the concerned communities, scholars, and stakeholders in the design of well-suited strategies to solve the numerous issues faced in the region.

## 1.4 Scope of the Book

As discussed in previous sections, the main objective of this book is to highlight opportunities for making resilient and sustainable communities in arid and semi-arid regions of Mexico. For this, we organized the content into seven chapters, dealing from the peri-urban and urban regional scale to the materiality of planning strategies and eco-building. Their generalities are presented below.

Cities must provide decent and affordable dwellings, access to essential services, efficient transportation systems, green areas, and public spaces to achieve sustainability. Chapter 2 evaluates three semi-arid cities in Mexico with similar demographics and infrastructure. Natural conditions such as its topography, climate, and communication routes that shape cities and their containment polygons were analyzed. An adverse effect is a predominantly arid and semi-arid climate since the water resource is limited, even when there is a diversity of other natural resources (soil, vegetation). Some areas of Mexican cities have precarious dwellings and limited access to essential services, marginalizing the less-favored population. At the same time, local policies promote urban growth toward sectors focused on economic growth, neglecting the basic needs of its inhabitants, mainly in peri-urban areas. Physical factors such as topography play a relevant role in urban development since zones with steep slopes are more prone to landslides; simultaneously, it is more complicated to supply water to the population if there is no efficient hydraulic infrastructure. On the other hand, those areas with no slope are exposed to flooding, and so both sites are vulnerable to natural hazards. Accessibility to cities is determined by their peripheral communication routes that cross the city. A well-connected city favors the mobility of its inhabitants and, therefore, the equitable development of their economic activities. Well-consolidated areas have adequate coverage of services, and the population has low poverty rates. Peri-urban areas are poorly consolidated or lack this basic coverage, with high poverty rates. The authors determined that the poorly consolidated zones were located in the historical centers (with obsolete infrastructure and equipment lacking in maintenance and rehabilitation), while peri-urban regions were growing without orderly planning. They concluded that it is necessary to establish criteria and indicators that guide the creation and application of public policies for cities and communities to be inclusive, safe, resilient, and sustainable.

Urban and peri-urban zones could adapt through public policies and citizen participation to promote inclusive planning, which fosters access to affordable dwellings, equipment, and infrastructure, especially for the less-favored population. Chapter 3 presents an urban sustainability index in a semi-arid city in central Mexico, making it easier to identify the degree of sustainability (or the lack of it) in the containment polygons according to their degrees of consolidation. This index considers social inclusion, decent dwelling, security, integration of vulnerable groups, and cultural and natural recreational spaces. In addition, it considers factors such as population density and the diversity of economic activities, which promote more efficient and lower-cost services and access to employment. The low-income and informal dwellings sectors show poverty reflected in limited access to land ownership, poorly built environment, high urban density, and limited access to essential services. These zones are in the peri-urban area and downtown, where the population has low resources and low education. In addition, the infrastructure and basic services still need to be available. It is desirable to increase living conditions with a focus on the physical state of access to essential services in the dwelling, prevent the deterioration of infrastructure and equipment, sociodemographic distribution, and the urban environment. Authors conclude that some of the improvements in the living conditions of society are the preservation of the historical and natural heritage since it provides a sense of belonging and willingness to take care of the local resources. The previous conclusion could be achieved by designing public policies involving societal participation.

As mentioned earlier, urban sustainability relies on its infrastructure and the preservation of its local ecosystem services. Chapter 4 revalues the historical stages of a railroad that, for a long time, had facilitated the activities of a booming mining town in central Mexico. The rugged topography defined the conditions of a narrow corridor that communicated toward the existing mines for transporting goods. In the urban area, old railway station facilities were partially transformed into public spaces while the remaining are still preserved. The authors revalued the current historical heritage of this aging infrastructure, as well as the physical conditions of the natural corridor. They evaluated the climate, vegetation, local geology, existing relics, and the perspective of the local indigenous population. The results of this work recommend reconverting the railway route analyzed, from the city's historic center to the peri-urban area, through a natural tourist corridor. The semi-arid climate and the transitional vegetation that goes from shrubs to conifers show the relevance of this area around the cultural and natural landscape, suitable for outdoor activities during certain seasons of the year. The authors concluded that there is an opportunity to recover the sustainability of this route by strengthening economic, social, and environmental links in a positive way as a planning strategy for the cultural and natural development of the area, preserving the ecosystem services that this natural corridor provides.

The traditional knowledge of the local people involves their longstanding knowledge and philosophy regarding their interaction with their natural environment. Throughout history, different social groups developed systems to adapt to various environmental challenges and have passed from generation to generation until today. In Mexico, since the Mexica, the humid hydraulic infrastructure known as chinampas (floating gardens) has been used, built on small islands adapted to areas susceptible to flooding during the rainy season. They still provide crops as an economic activity to the population throughout the year. There are also examples of dry hydraulic infrastructure, built as underground aqueducts that collected groundwater and transported it to the population and their agricultural zones in areas such as Iran and Iraq. Examples such as the above have provided water security since ancient times thanks to the management of storing and delivering water through local adaptations. Chapter 5 presents a reflection of experiences in countries of the Global South, such as Mexico, where there are severe problems of drought and water scarcity. Here are examples of the efforts made worldwide and over time to preserve the ecosystems, especially the services provided by the green and blue infrastructure (BGI) associated with water security. The authors explain that this type of infrastructure consists of an integrated system between the planning of the natural landscape and the urban environment through their interconnection. BGI adapts to reintroduce nature into built environments and help mitigate the adverse effects of climate change. Also, BGI is an alternative to improve the health of the less-favored people without access to health services. The BGI concept recognizes natural ecosystems as valuable as conventional infrastructure in the built space. It can be adapted to rural/urban transition zones, built/unbuilt areas, and natural/artificial areas. One of the most valuable things to underline is that the experience of local indigenous societies still shows us that it is possible to continue incorporating these sustainable infrastructure practices to mitigate the climate change effects and to preserve natural resources for future generations.

Another known example of using natural resources through the knowledge of the local indigenous population in the Global South is eco-building with materials such as adobe and rammed earth. Chapter 6 presents that among the main benefits of this eco-building practice is using locally available natural resources such as soil. Building with local soil reduces the extraction of construction materials such as cement, gravel, or sand, whose adverse environmental effects are widely documented. Also, they are an affordable alternative for the less-favored population that requires adequate and decent dwellings in urban and peri-urban areas. The authors highlight the advantages of this eco-building process, particularly in regions with arid and semi-arid climates, where homes are required to provide their inhabitants with comfort in the face of sudden changes in temperature and safety from natural disasters, such as hurricanes and earthquakes. There are pre-Hispanic antecedents of its use in different parts of Mexico. The local indigenous population used adobe and rammed earth and reinforced them with nopal or corn fibers for the inclement weather because they were locally available. This eco-building practice resumes its importance by incorporating other materials to improve their thermal, acoustic, and durability properties. In addition, using these materials positively impacts the environment since it reduces fuel consumption for transportation by being available locally. Also, incorporating waste produced by the population contributes to reducing the use of virgin or synthetic materials and to improving their properties. Such is the case of using palm fibers, hemp, low-quality wool, recycled polyethylene from fishing nets, and bird feathers, among others. These examples also demonstrate the possibility of incorporating circular economy practices that support the local economy by introducing waste from different local activities into a positive cycle to protect the less-favored population living in unsustainable conditions.

Finally, one of the main engines of a country's economy is its road infrastructure. Goods and people move through it, requiring safe, durable, and low-maintenance roads. Mexico's north and central zones concentrate a large part of the economic activity and population at the national level. Chapter 7 evaluates this area where the arid and semi-arid climate challenges the quality of such an important infrastructure. In countries of the Global South, such as Mexico, it is essential to consider sustainable alternatives that help develop innovations to reconvert their road infrastructure into reliable and quality systems since their expansion is necessary. Technological innovations incorporate waste to improve the properties of asphalt, one of the main materials used to pave roads. Waste such as rubber and other plastic derivatives have proven to be suitable for modifying this material because they share characteristics that make them compatible with asphalt. There is data on the use of shredded tire waste that has increased performance, mechanical properties, service life, and reduced highway construction and maintenance costs. Another example is using polymers obtained from recycled plastics as a substitute for virgin aggregates, geomembranes or to modify the binder. Many studies support favorable results in asphalt pavements modified with these residues. Among others, reduced deformations and distortions, greater tensile strength, and increased rigidity have been achieved. The authors conclude that some areas of opportunity in Mexico are through the regulated reuse and recycling of plastics. At the same time, they are an eco-building practice that encourages substituting natural resources with waste to mitigate its negative environmental impact.

In conclusion, this book establishes proposals to adapt the livable spaces of Mexico placed in arid and semi-arid regions from a resilient and sustainable perspective. Each chapter proposes recommendations focused on adapting the Sustainable Development Goals (SDGs) to help improve the quality of life of its population and preserve both natural and cultural resources. The scope of this work may be helpful in other areas with similar climates in the Global South.

## References

Andrić I, Koc M, Al-Ghamdi SG (2019) A review of climate change implications for the built environment: impacts, mitigation measures and associated challenges in developed and developing countries. J Clean Prod 211:83–102

Bamigboye GO, Bassey DE, Olukanni DO, Ngene BU, Adegoke D, Odetoyan AO, Kareem MA, Enabulele DO, Nworgu AT (2021) Waste materials in highway applications: an overview on generation and utilization implications on sustainability. J Clean Prod 283:124581

Beier P, Noss RF (1998) Do habitat corridors provide connectivity? Conserv Biol 12(6):1241–1252

Bellezoni RA, Meng F, He P, Seto KC (2021) Understanding and conceptualizing how urban green and blue infrastructure affects the food, water, and energy nexus: a synthesis of the literature. J Cleaner Prod 289:125825

Bisht A (2021) Conceptualizing sand extractivism: deconstructing an emerging resource frontier. Extr Ind Soc 8(2):100904

Buck KD, Summers JK, Smith LM (2021) Investigating the relationship between environmental quality, socio-spatial segregation and the social dimension of sustainability in US urban areas. Sustain Cities Soc 67:102732

Budds D (2019) How do buildings contribute to climate change? https://archive.curbed.com/2019/9/19/20874234/buildings-carbon-emissions-climate-change. Accessed 13 May 2023

Campbell S (1996) Green cities, growing cities, just cities? J Am Plann Assoc 62(3):296

Castells M (ed) (2002) Livable cities? Urban struggles for livelihood and sustainability. University of California Press, Berkeley

Charai M, Mezrhab A, Moga L (2022) A structural wall incorporating biosourced earth for summer thermal comfort improvement: hygrothermal characterization and building simulation using calibrated PMV-PPD model. Build Environ 212:108842

Clarke M (2018) Global South: what does it mean and why use the term. Global South Political Commentaries 8

ConstructionPlacements (2021) What is the future of construction jobs? https://www.constructionplacements.com/future-of-construction-jobs/. Accessed 29 May 2022

Crutzen P (2002) Geology of mankind. Nature 415:23

Cueva J, Yakouchenkova IA, Fröhlich K, Dermann AF, Dermann F, Köhler M, Grossmann J, Meier W, Bauhus J, Schröder D, Sardemann G, Thomas C, Romero-Carnicero A, Saha S (2022) Synergies and trade-offs in ecosystem services from urban and peri-urban forests and their implication to sustainable city design and planning. Sustain Cities Soc 82:103903

D'Odorico P, Porporato A, Runyan C (2019) Dryland ecohydrology, 2nd edn. Springer, Cham, pp 2–4

Dados N, Connell R (2012) The global south. Contexts 11(1):12–13

De Sousa CA (2003) Turning brownfields into green space in the City of Toronto. Landsc Urban Plan 62(4):181–198

Dixon W (2010) The impacts of construction and the built environment. Briefing Note 33

Easton S, Lees L, Hubbard P, Tate N (2020) Measuring and mapping displacement: the problem of quantification in the battle against gentrification. Urban Stud 57(2):286–306

Elhacham E, Ben-Uri L, Grozovski J, Bar-On Y, Milo R (2020) Global human-made mass exceeds all living biomass. Nature 588(7838):442–444

Elmqvist T, Andersson E, Frantzeskaki N, McPhearson T, Olsson P, Gaffney O, Takeuchi K, Folke C (2019) Sustainability and resilience for transformation in the urban century. Nature Sustain 2(4):267–273

Elmqvist T, Andersson E, McPhearson T, Bai X, Bettencourt L, Brondizio E, Colding J, Daily G, Folke C, Grimm N, Haase D (2021) Urbanization in and for the anthropocene. Urban Sustain 1(1):1–6

Farina A, Zanetti MC, Santagata E, Blengini GA (2016) Life cycle assessment applied to bituminous mixtures containing recycled materials: crumb rubber and reclaimed asphalt pavement. Resour Conserv Recycl 117:204–212

Fernandes J, Peixoto M, Mateus R, Gervásio H (2019) Life cycle analysis of environmental impacts of earthen materials in the Portuguese context: rammed earth and compressed earth blocks. J Cleaner Prod 241:118286

Foundation BNP Paribas (2021) Carbon story. Global Carbon Atlas and the Institut Pierre Simon Laplace (IPSL), France. http://www.globalcarbonatlas.org/es/outreach. Accessed 10 Sep 2022

Godschalk DR (2004) Land use planning challenges: coping with conflicts in visions of sustainable development and livable communities. J Am Plan Assoc 70(1):5–13

Hassan KE, Elghali L, Soerby CR (2004) Development of new materials for secondary and recycled aggregates in highway infrastructure. TRL Report TRL598

Hernández A, Miranda M, Arellano EC, Saura S, Ovalle C (2015) Landscape dynamics and their effect on the functional connectivity of a Mediterranean landscape in Chile. Ecol Indic 48:198–206

Huber-Sannwald E, Ribeiro Palacios M, Arredondo Moreno JT, Braasch M, Martinez Pena RM, de Alba Verduzco JG, Monzalvo SK (2012) Navigating challenges and opportunities of land degradation and sustainable livelihood development in dryland social–ecological systems: a case study from Mexico. Philos Trans R Soc B Biol Sci 367(1606):3158–3177

Hürlimann AC, Nielsen J, Moosavi S, Bush J, Warren-Myers G, March A (2022) Climate change preparedness across sectors of the built environment—a review of literature. Environ Sci Policy 128:277–289

IPCC (2018) Global warming of 1.5°C. An IPCC special report on the impacts of global warming of 1.5°C above pre-industrial levels and related global greenhouse gas emission pathways, in the context of strengthening the global response to the threat of climate change, sustainable development, and efforts to eradicate poverty. https://www.ipcc.ch/sr15/. Accessed 3 Aug 2022

Jim CY, Chen WY (2006) Recreation–amenity use and contingent valuation of urban greenspaces in Guangzhou. China. Landsc Urban Plan 75(1–2):81–96

Liu Z, Xu C, Xu T, Jia H, Zhang X, Chen Z, Yin D (2021) Integrating socioecological indexes in multiobjective intelligent optimization of green-grey coupled infrastructures. Resour Conserv Recycl 174:105801

Macháč J, Brabec J, Arnberger A (2022) Exploring public preferences and preference heterogeneity for green and blue infrastructure in urban green spaces. Urban for Urban Green 75:127695

Market Prospects (2021) Global construction trends. https://www.market-prospects.com/articles/global-construction-industry-trends. Accessed 6 Jun 2022

McGrath M (2019) One million threatened species, the worrying UN report on the impact of humans on the planet. BBC News World Online. https://www.bbc.com/mundo/noticias-48176068. Accessed 3 Aug 2022

Melo-dos-Santos S, Fernandes-Ferreira-Silva J, Chaves-dos-Santos G, Martins-Torres de Macedo P, Gavazza S (2019) Integrating conventional and green roofs for mitigating thermal discomfort and water scarcity in urban areas. J Clean Prod 219:639–648

Miller N (2021) The industry creating a third of the world's waste. BBC Future Planet. https://www.bbc.com/future/article/20211215-the-buildings-made-from-rubbish. Accessed 23 Jun 2022

Monteiro PJ, Miller SA, Horvath A (2017) Towards sustainable concrete. Nat Mat 16(7):698–699

Moran MR, Bilal U, Dronova I, Ju Y, Gouveia N, Teixeira-Caiaffa W, de Lima-Friche AA, Moore K, Miranda JJ, Rodríguez DA (2021) The equigenic effect of greenness on the association between education with life expectancy and mortality in 28 large Latin American cities. Health Place 72:102703

Nagendra Nagendra H, Bai X, Brondizio ES, Lwasa S (2018) The urban south and the predicament of global sustainability. Nat Sustain 1:341–349

Opp SM (2017) The forgotten pillar: a definition for the measurement of social sustainability in American cities. Local Environ 22(3):286–305

Ortiz-Hernández J, Lucho-Constantino C, Lizárraga-Mendiola L, Beltrán-Hernández I, Coronel-Olivares C, Vázquez-Rodríguez C (2016) Quality of urban runoff in wet and dry seasons: a case study in a semi-arid zone. Environ Sci Pollut Res 23:25156–25168

Pannunzio-Ribeiro M, de Mello K, Valente RA (2022) How can forest fragments support protected areas connectivity in an urban landscape in Brazil? Urban for Urban Green 74:127683

Paterson W, Rushforth R, Ruddell BL, Konar M, Ahams IC, Gironás J, Mijic A, Mejia A (2015) Water footprint of cities: a review and suggestions for future research. Sustainability 7(7):8461–8490

Peduzzi P (2014) Sand, rarer than one thinks. Environ Dev 11:208–218

Sampson RJ (2017) Urban Sustainability in an age of enduring inequalities: advancing theory and ecometrics for the 21st-century city. Proc Natl Acad Sci USA 114(34):8957–8962

Shackleton CM, Cilliers SS, du Toit MJ, Davoren E (2021) The need for an urban ecology of the Global South. In: Shackleton CM, Cilliers SS, Davoren E, du Toit MJ (eds) Urban ecology in the global South. Springer, Cham, pp 1–26

Sneddon C, Howarth RB, Norgaard RB (2006) Sustainable development in a post-Brundtland world. Ecol Econ 57(2):253–268

Spinoni J, Barbosa P, Cherlet M, Forzieri G, McCormick N, Naumann G, Vogt JV, Dosio A (2021) How will the progressive global increase of arid areas affect population and land-use in the 21st century? Glob Planet Change 205:103597

Steffen W, Sanderson RA, Tyson PD, Jäger J, Matson PA, Moore B III, Oldfield F, Richardson K, Schellnhuber HJ, Turner BL, Wasson RJ (2004) Global change and the earth system: a planet under pressure. Springer, Heidelberg

Torres A, Simoni MU, Keiding JK, Müller DB, Zu Ermgassen SOSE, Liu J, Jaeger JAG, Winter M, Lambin EF (2021) Sustainability of the global sand system in the Anthropocene. One Earth 4(5):639–650
UN (2019) World Urbanization Prospects: The 2018 Revision (ST/ESA/SER.A/420) United Nations, Department of Economics and Social Affairs, Population Division
WHC (2017) Guidelines for the implementation of the world heritage convention. https://whc.unesco.org/en/guidelines/. Accessed 30 Aug 2022
WMO (2022) United in science 2022: a multi-organization high-level compilation of the most recent science related to climate change, impacts and responses. World Meteorol Organ
Woudstra J, Fieldhouse K (eds) (2000) The regeneration of public parks. Taylor and Francis, London
Zalasiewicz J, Waters CN, Williams M, Barnosky AD, Cearreta A, Crutzen P, Ellis E, Ellis MA, Fairchild IJ, Grinevald J, Haff PK (2015) When did the anthropocene begin? A mid-twentieth century boundary level is stratigraphically optimal. Quat Int 383:196–203
Zhang R, Peng S, Sun F, Deng L, Che Y (2022) Assessing the social equity of urban parks: an improved index integrating multiple quality dimensions and service accessibility. Cities 129:103839

# Chapter 2
# Sustainable Cities and Urban Morphology in Arid and Semi-Arid Climates of Mexico

**Alma Delia Juárez-Sedano and Miguel Ángel González-Loza**

**Abstract** This chapter addresses the urban form of three Mexican cities in areas with arid and semi-arid climates, considering the Sustainable Development Goals (SDGs) established in the 2030 Agenda. The aim is to analyze the urban form of cities and its relationship with the targets set in SDG 11, which refers to cities and communities to make them inclusive, safe, resilient, and sustainable. For this, three Mexican cities with similar characteristics were compared: Querétaro, San Luis Potosí, and León, with 1.59, 1.27, and 1.9 million inhabitants, respectively. The elements considered for this analysis were natural conditions where they are located, the routes that give origin and shape to the city, and their containment polygons. With the above, it was possible to identify that the form of cities plays a vital role in a town being inclusive, resilient, and sustainable by determining the physical order that directly affects the accessibility and, therefore, the physical structure of the city, which in turn governs the practical use of resources.

**Keywords** Path matrix · Containment perimeters · Urban poverty · Inclusive cities · Accessibility

A. D. Juárez-Sedano (✉)
Engineering and Architecture Academic Area, Autonomous University of Hidalgo State, Carr. Pachuca-Tulancingo Km 4.5, Col. Carboneras, Mineral de La Reforma, C.P. 42184 Hidalgo, Mexico
e-mail: ajuarez@uaeh.edu.mx

M. Á. González-Loza
Higher Public Normal School of the State of Hidalgo, Av Pirules 213, El Tezontle Extension, C.P. 42080 Pachuca de Soto, Hgo, Mexico
e-mail: loza_miguel@ensupeh.edu.mx



L. Lizárraga-Mendiola et al. (eds.), *Sustainable Spaces in Arid and Semiarid Zones of Mexico*, The Latin American Studies Book Series,
https://doi.org/10.1007/978-3-031-53418-8_2

## 2.1 Sustainable Cities and Communities

Urban sustainability is a challenge that involves spatial and temporal scales (Sobrino et al. 2015). It constitutes one of the principles regulating how cities grow and develop. It is also a response to the social and environmental imbalance in the world. As a premise, when creating a city, urban sustainability must lay the foundations for thinking about and designing it since the problems that cities face today result from excessive growth, improvisations, poor organization, poor design, and witty administrations.

In this sense, Sustainable Development Goal (SDG) 11, established in the Sustainable Development Agenda of the United Nations, seeks to make cities more inclusive, safe, resilient, and sustainable by giving governments the responsibility to generate the conditions for these premises to occur within a framework of regulation, administration, and investment. Thus, the search for urban sustainability, as an agreement between nations in the process of making cities, requires governments to apply policies that give direction to the growth of population centers in each country and the world. This implies that local governments lay the foundations so that the inhabitants of cities and communities have decent housing, essential services, efficient transportation systems, green areas, and public spaces. In the design of spaces, inclusiveness and security regarding natural disasters should always be considered in areas whose conditions are determined by the location and physical morphology of the territory.

Urban sustainability is a benchmark for community life in balance. In contrast, inequality opens the gap between those who enjoy a comfortable life and those who suffer from social deprivation. The Spatial Justice Theory (Lefevbre 1968; Harvey 1988; Soja 1996) and the search for urban sustainability emphasize balance as a strategic part of city planning.

Some cities are favored over others because they are located near natural resources and have non-extreme climates. The same happens with certain areas within cities close to services and infrastructure. A good location allows cities to be more competitive, and, for citizens, it leads to better opportunities to develop since it implies lower costs (time and money). Energy use, derived from extreme temperatures, is closely related to the position of a place concerning the distance to the Equator, which impacts the economy of households and companies to the extent that they seek to inhabit comfortable areas.

Taking into account that the most sustainable cities in the world owe it to dense, compact growth (Campos-Sánchez et al. 2018), adequate use of resources, and the implementation of public policies in favor of sustainable development and investments, the planning of cities must be oriented to each of these conditions. Compact cities generate proximity and accessibility and include social and environmental aspects (Naess 2014). Using resources implies both those derived from natural resources in their qualitative elements and those from their processes (Vargas et al. 2017). Thus, these cities have green spaces, accessibility, efficient transportation, sustainable infrastructure, and economic affordability.

Cities such as London, Edinburgh, Stockholm, and Singapore stand out for the investments made in caring for the environment through the optimization of transport and sustainable infrastructure, as well as investments that stimulate economic affordability and result in more job opportunities. Among the recommendations suggested by Honnie and Grimm (2020) for environmental justice in planning urban green spaces are to seek equitable access, listen to all population groups, and generate interaction and security within the areas. For this reason, it is necessary to develop, with the help of the planning of the territory and resources, a balance not only environmental but also economic, political, and social that has a fair impact on all citizens, given that the growth of cities is imminent.

Year after year, the number of cities grows, as does their population. In 2020, 1,934 metropolises with more than 300,000 inhabitants were identified (UN-Habitat 2020). The great challenge facing urban concentrations today is understanding the growth model to follow based on the same environment where the city is located.

According to the National Urban System, there are 401 cities in Mexico inhabited by 92.7 million people (SEDATU 2018). From these, 74 are metropolitan areas, 132 are conurbations, and 195 are categorized as urban centers. Between 2010 and 2015, metropolitan areas increased from 59 to 74; 36 new cities, seven conurbations, and 29 urban centers were added. Given that urban concentrations continue growing, better resource and space use is imperative. For this, the organization and form of growth of urban centers must correspond to the natural conditions where they are located.

In Mexico, seven climates have been identified (SEMARNAT 2022): (1) warm subhumid, (2) arid and semi-arid, (3) very arid or arid desert, (4) warm humid, (5) temperate subhumid, (6) temperate humid, and (7) cold, of which the first five predominate. The most important factors that determine the type of climate are geographical position, distance from the Equator and the sea, and the land's relief. The natural characteristics of the territory play a preponderant role in the use and design of cities, taking into account that even the elements of the towns' morphology modify the behavior of urban temperatures (Castro-Conrado et al. 2015).

Cities and communities with a view to sustainability must consider their design based on the natural characteristics of their location since climates determine the temperature and the resources in it. The topography establishes the configuration of urban forms, as it generates, from the use of slopes and better use of resources such as water, ventilation, and lighting. Therefore, one cannot think of making better use of these resources if one does not identify the conditions, advantages, and disadvantages that these locations and the resources around them are generating.

The present work is a comparative study of three Mexican cities, Querétaro, San Luis Potosí, and León. These were chosen for their similarities: population, temperate arid and warm semi-arid climates, and a topography characterized by mountains and hills. The topics addressed in this analysis are the natural conditions of the territory, such as the climate, the topography, and its slopes; urban structure based on its main roads and their role in shaping the city; and the relationship between containment polygons and urban poverty. The main purpose is to assess to what extent the form and structure of these three cities favor achieving the targets established as part of SDG 11.

## 2.2 Physical and Natural Characteristics of the Cities

The constant and rapid population movements and some well-off social groups have not allowed the construction of a scheme to design cities in a balanced way, marginalizing those who have the least to live in remote and unsuitable areas to inhabit (Rubalcava and Schteingart 2012). Thus, the housing deficit has characterized the peripheries of Mexican cities with precarious housing and limited access to services (González-Arellano 2011), neglecting areas within the cities due to the disinterest of the rulers and marginalizing the less-favored.

Cities, the columns of national economies (Castro-Castro 2018), compete with each other. In this competition, they tend to grow by absorbing places, suitable and not suitable for inhabiting through production relations. These have served as an impetus to strengthen the economies within and, in part, to justify the investment and appropriation of some areas in the cities. Relations that once structured cities become networks that generate territorial divisions (Ramírez 2010) while the impulse is not given equitably and according to the conditions and needs of the territory and the inhabitants, generating territorial disparities that remain visible and are more marked on the periphery.

The physical and natural characteristics of the areas where cities are located determine whether or not a city may be more sustainable. Regardless of the socio-environmental situation of the people, living in the peri-urban conditions the accessibility of the people. As they are recently-created areas, they do not have services and infrastructure since each geographical location has its challenges (Linares-García and Hernández-Quirama 2018).

The planning and urban design of cities, as a permanent exercise, if it considers the elements in the natural environment, could lead the sites in the territory, in the near future, to sustainable cities.

### *2.2.1 Generalities*

According to the classification of Vidal-Zepeda (2005), the Mexican cities to be analyzed are part of the 04-north region. This region is characterized by aridity and low rainfall, high temperatures during the day in summer, and low temperatures in winter due to elevation, low pressure, and isolated winds (Castro-Conrado et al. 2015). Querétaro, San Luis Potosí, and León are located in the north of Mexico City, as shown in Fig. 2.1.

This region is located within a larger one, delimited by the most populated cities in the country (Mexico City, 21.34 million inhabitants; Monterrey, 4.48; Guadalajara, 4.8; and Puebla, 2.96). The municipalities that make up the study metropolitan areas are Apaseo el Alto, Corregidora, Huimilpan, El Marqués, and Querétaro, in Querétaro; San Luis Potosí, Soledad de Graciano Sánchez and Zaragoza, in San Luis Potosí; and León and Silao de la Victoria, in León. Their population, in millions of

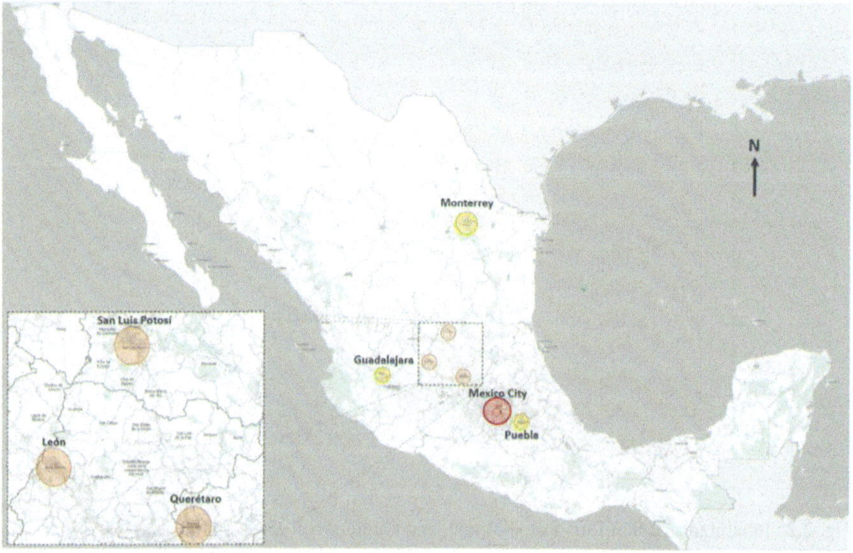


**Fig. 2.1** Location of the study cities in Mexico (INEGI 2022)

inhabitants, is 1.59, 1.27, and 1.9, respectively. Their demographic growth between the 2010 and 2020 censuses was 37.3% for Querétaro, 19.4% for San Luis Potosí, and 19.6% for León (Data-México 2020). When comparing these growth rates with that of Mexico City (8.39%; Data-México 2020) and the national average (1.2%; Saldívar and Badillo 2021), it was observed that the study areas had a high growth in the last decade, which demanded the generation of a significant number of services and infrastructure in a short time.

Since 1945, the population growth trend of the three cities has been upward (Fig. 2.2). Above all, in the case of Querétaro, it went from having 49,000 inhabitants to 95,000 in 2005, doubling its population and surpassing that of San Luis Potosí. By 2020, its population exceeded 1.5 million, representing a 15-fold growth in just 15 years. Thus, at the end of the twentieth century, the national urban system left its monocentric configuration to become polycentric by forming nine metropolises with more than one million inhabitants (Garza-Villarreal 2002).

In these cities, a topography configured by mountains and hills was identified. Places with wide biodiversity provide vegetation, oxygen, fauna, minerals, and great views. At the same time, the natural edges contain the urban sites in some areas of the cities. The settlements are concentrated on slopes below 6% (Fig. 2.3).

According to Higueras (1998), when determining land uses, topography gives the guidelines to consider those areas suitable for inhabitation. For his part, Bazant (1984) states that on slopes greater than 15%, the land is subject to significant erosion and landslides. Urbanization costs in areas whose slopes exceed 15% become expensive at the time of urbanization. For example, incorporating a drainage system in steep slopes involves considering drop wells that break with slope speeds or speed breaker

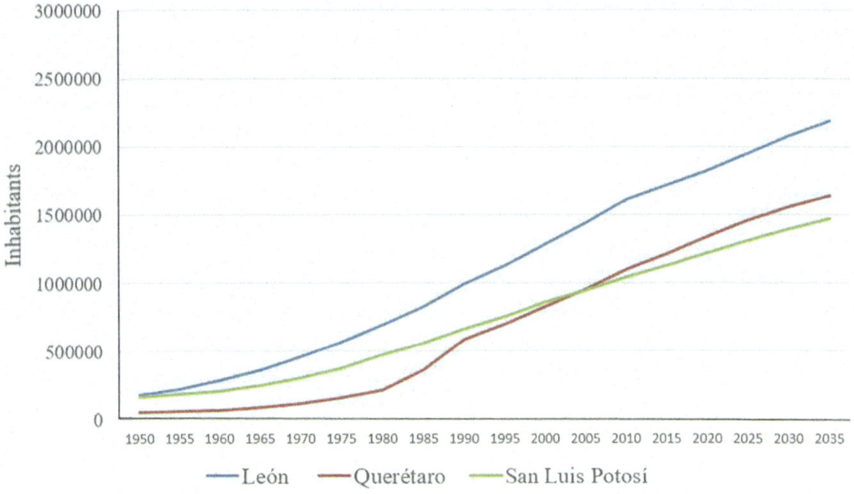


**Fig. 2.2** Inhabitants of San Luis Potosí, León, and Querétaro, 1950–2035 (UN 2018)

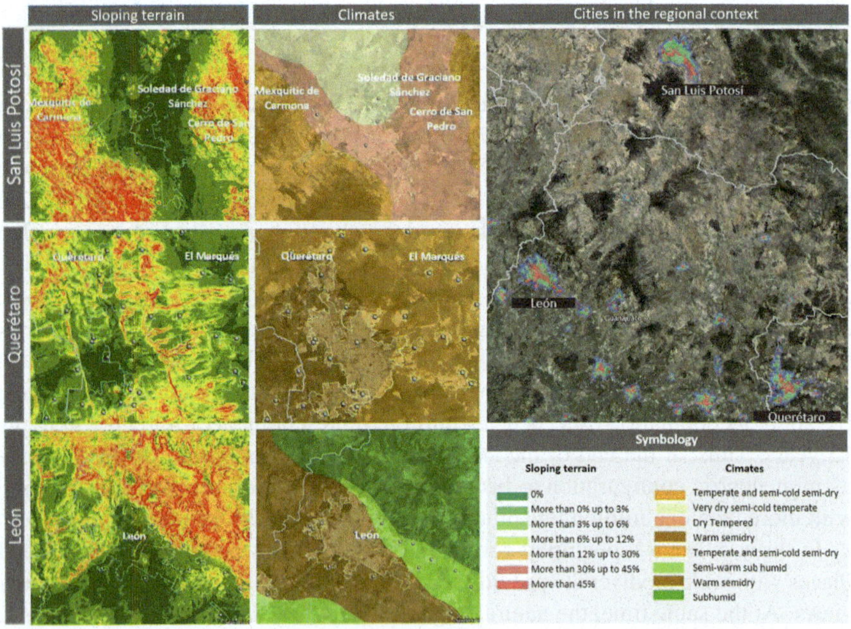


**Fig. 2.3** Physical conditions by study city (INEGI 2022)

boxes. In the case of the road system, the safety aspects also result in high costs since the roads must be lengthened to generate reduced slopes and to develop comfort, safety, and accessibility when transiting.

In an examination of the slopes and the location of the settlements in the cities of analysis, it was identified that the city of Querétaro presents a rugged topography in the northeast and southeast areas; León in the northeast area; and San Luis Potosí in the northeastern and southwestern areas. In the three cities, this limits the locations of the settlements to less rugged areas, given that in the rugged areas, the slopes range between 12 and 45% (Fig. 2.3).

### 2.2.2 *Brief Description of the Physiography*

San Luis Potosí. It presents a topography determined by two mountain systems whose slopes, at their highest points, are above 45%, confining the city's growth in its extreme southwest and northeast points, connecting the north to Monterrey and the south with the cities of León and Aguascalientes. Three climates prevail in this city: arid temperate, very arid, semi-cold, and warm and semi-arid, from which it is possible to recommend recreational use (parks or wetlands), for the low-lying area, subject to flooding. It has a 0% slope that is considered a flood zone, while slopes at 3% promote a better functioning of the drainage system and a lower investment cost.

Querétaro. Unlike San Luis Potosí and León, it has a territorial shape of hills whose slopes range between 6 and 30%. Its center is surrounded by flat edges, leading to flooding and damages in settlements in the rainy season. Steep slopes allow excellent views. Its location has been configured mainly to the north, with a significant branch that goes to the southeast, connecting with Tequisquiapan and San Juan del Río. To the north, it borders San Luis Potosí and San Miguel de Allende and, to the southwest, with Celaya and Morelia. The city's climate, being warm semi-arid and temperate semi-cold semi-arid, generates good ventilation and pleasant warmth.

León. It has an elongated shape that goes from the southeast to the northwest. Limited to the northeast, to a greater extent, by a mountainous area whose slopes range between 6, 12, 30, and up to 45%. This city presents more regular emplacements in areas whose inclinations are 0, 3, and 6% to the city's south. Due to its slopes, there is a pattern of growth to the west, an area that connects with cities such as Guadalajara and Aguascalientes, and to the southeast, where there are cities like Irapuato, Guanajuato, Celaya, San Miguel de Allende, Morelia, and Querétaro. This last area is subject to flooding, as it is in the lower part of the city. The mountainous region of the northeast contains the growth of settlements and allows their location only on the slopes. The climates in the city are warm semi-arid, semi-warm subhumid, and subhumid, which mainly provide more natural resources in the northeast.

## 2.3 Routes Connecting the Cities

The origin of the city's shape is the street. The shape of the city routes is determined by the destinations to which the inhabitant is directed. Since accessibility is one of the criteria used to measure sustainability in cities and communication routes and transport systems determine it, it is necessary to analyze the routes shaping cities.

According to Caniggia (1995), cities are configured from routes in his studies on the city's topology. The same derives from a matrix path. This is defined as the first significant road in the city, which becomes the backbone of the road system. They are followed by the implantation routes that, in the cases of Mexican cities, are those at the second level of hierarchy that generally crosses the main route, connecting with each other those areas built as an expansion and, later, the periphery of the cities. These road connections can be in the form of rings or axes. Finally, the implant paths are connected with union paths. Through this, we access the neighborhoods that make up the city.

In the case of the three study cities, the main routes shape them since they connect with the immediate cities of higher hierarchy, as follows. For Querétaro, Paseo de la República Avenue, which continues the Mexico-Querétaro highway, borders the city of Querétaro with Mexico City to the south and the city of San Luis Potosí to the north. San Luis Potosí's main route is Paseo de la República Avenue, which connects to the south with the city of Querétaro and to the north with Saltillo and Monterrey. While Adolfo López Mateos Boulevard, connecting with Silao and Irapuato to the southeast, and Lagos de Moreno and Aguascalientes, to the northwest, is the main route of León. These main paths are crossed by peripheral rings, such as the Libramiento sur Poniente Avenue in Querétaro, the Peripheral Ring in San Luis Potosí, and José María Morelos Boulevard, in León (Fig. 2.4).

The configuration and design of the matrix paths in León and San Luis Potosí have been based on a territorial form whose sloping terrain has not hindered its layout, as it has not exceeded 15%. In the case of Querétaro, the territory's shape is uneven, which has made road layout and connections difficult, especially in areas with steeper slopes. This leads to less accessibility compared to the other two cities, in the strict sense of the main routes.

Concerning the implantation paths, which are the city's ramifications and extensions, they determine the widths and, therefore, the very form of the cities. In addition, they establish unions between neighborhoods and other urban localities, allowing the consolidation of areas within the city that do not precisely derive from downtown, contrary to the usual trend. The urban containment perimeters are presented in Fig. 2.4.

The routes of the Querétaro and San Luis Potosí cities have the central axis fed by an implantation route. In the case of San Luis Potosí, it is an internal circuit. In the case of Querétaro, the main implantation route drifts in two directions but under the same line. In contrast, León is configured from an implantation path that results in ramifications. These give off other ways at different times. In turn, the main route

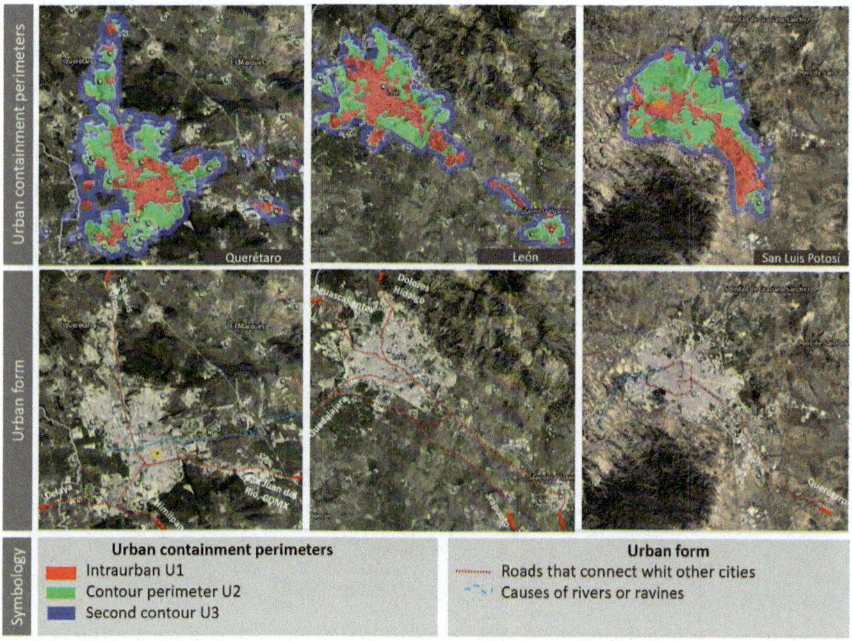


**Fig. 2.4** Urban form from the routes and urban consolidation, contention, and expansion perimeters (INEGI 2022)

of León has a parallel alternative road that runs through the city, but it is connected in two moments through implantation routes.

In this sense and according to the data provided by the Mexican Institute for Competitiveness (IMCO 2021), it was possible to identify León as the city with the best accessibility and infrastructure functioning (Fig. 2.5), given the way its routes cross the city from different points. Beyond the data provided in IMCO (2021), the urban form as a product of the routes shows a configuration that has resulted in better connectivity inside and outside cities (Fig. 2.5).

In Fig. 2.6, it can be seen that more than 25% of the inhabitants in these three cities spend between 30 min and one hour commuting to their workplaces while, on average, 7.5% can travel between one and two hours. The national average shows that 30% of the population spends up to 15 min commuting to their workplace, and 26% between 16 and 30 min (Data-México 2020). It may be argued that a large part of the time spent on transfers is due to the design of the streets and the distance of the routes, the urban form that cities have, and the proximity of services for residential areas. In that case, they play a predominant role in saving time. This implies an equitable distribution of services and a better design of routes.

In Mexico, there are support instruments to design roads, such as the Manual of Road Design for Mexican Cities (Aguirre-Benítez 2019), which addresses the design of streets from four principles: safety, inclusion, resilience, and sustainability. In

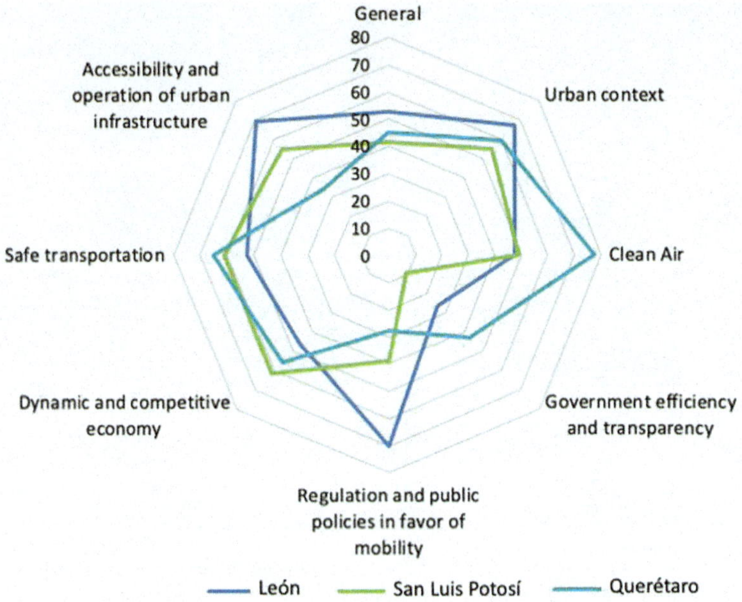


**Fig. 2.5** Urban mobility: better connected neighborhoods for more inclusive cities (IMCO 2021)

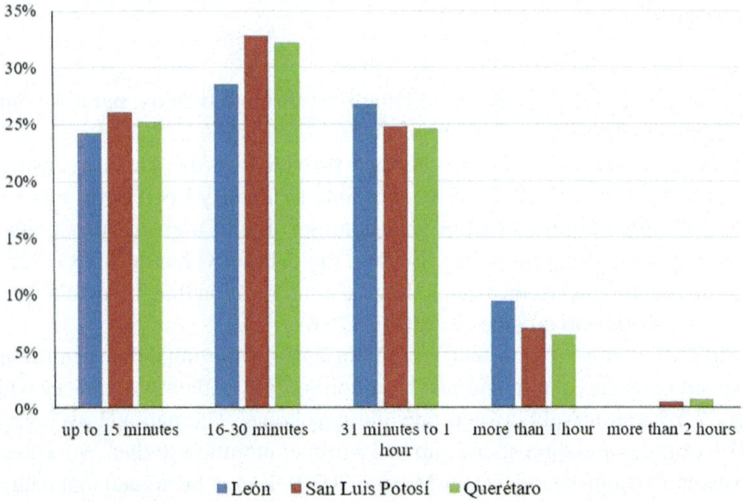


**Fig. 2.6** Percentage of inhabitants according to commuting time to work, by city, in 2020 (Data-México 2020)

addition, this manual has been designed based on priorities and hierarchies according to the type of mobility. In this hierarchy, the pedestrian is at the top, followed by cyclists, public transport and emergency vehicles, cargo transport, and, in the end, private cars.

## 2.4 Containment Polygons

The territory in the cities has had to be classified for its understanding and use according to the attributes with which it has been endowed (natural and artificial). Government institutions have taken on the task of organizing the territory and have established zoning within cities to define territorial limits with which to work.

The Secretariat of Territorial and Urban Agrarian Development, through the National Housing Commission, has built urban containment polygons to determine the areas that are close to services in Mexican cities registered within the National Urban System, creating a reference framework that allows identifying those areas within the cities that offer better living conditions. These polygons have been used as a reference to contrast the information that refers to the consolidated areas and urban poverty.

In this sense, the organization of the perimeters has been subject to three territorial limits. The first corresponds to the one that delimits the consolidated areas, called U1, which is characterized by having essential services (water and drainage), sources of employment, equipment (health, education, public spaces), and commercial areas. The second, called U2, corresponds to a delimitation of the areas within the city that only have essential services (water and drainage) at 75%. The third, U3, delimits those areas of urban expansion that still do not have all the essential services (Fig. 2.4).

When analyzing these containment polygons in the study cities, it was identified that they maintain the following configurations. Querétaro presents an irregular and fragmented consolidated area that predominates in the city's historic center, expanding to the north in an elongated way and in a fragmented way toward the southwest. In contrast, other smaller consolidated centers are located in the west, north, and southeast peripheral areas. For their part, the moderately consolidated areas have been configured more irregularly around the consolidated areas, which are directly connected to the expanding areas, except those found in the southwest. Thus, the consolidated areas are closely linked to the margin of the parent and implantation routes, generating a close relationship between the most important routes in the city and the consolidated areas.

León reveals an elongated shape configured around its parent path. It defines a location of the first perimeter that starts in the northwest area and ends intermittently in the southeast area. It sometimes widens toward the north and south of the city with minor ramifications, also due to the connections with Guadalajara and Dolores Hidalgo. The U3 containment polygons, defined by having a lower percentage of essential services, are areas immediately surrounding the consolidated urban containment polygons, especially and to a greater extent, to the south of the city.

For San Luis Potosí, two shapes can be seen: a circular shape to the north and a strip that emerges to the southeast. The circular shape presents a configuration where a consolidated area is located in the center, surrounded by a U2 area, which U3 also surrounds. The second form, the fringe, is characterized by a primarily consolidated area. To the south, there is a U3 zone, and to the east of the consolidated zone, a thin strip of U2, followed by another light strip of U3.

As already shown, there are tight areas of the city configured from the routes (Fig. 2.4). However, we still do not know the relationship between these tight areas and the location of urban poverty. As stated by Salinas-Arreortua and Soto-Delgado (2019), those consolidated places have been destined for the population with medium and high incomes, marginalizing those with low-income to growing areas and lacking services. Therefore, it is necessary to review poverty by basic statistical geographic areas (BSGA) in each of the three study cities and superimpose the contours of the containment polygons (Fig. 2.7), bearing in mind that the BSGA are subdivisions of the municipal territories that allow the organization and visualization of information based on smaller physical limits.

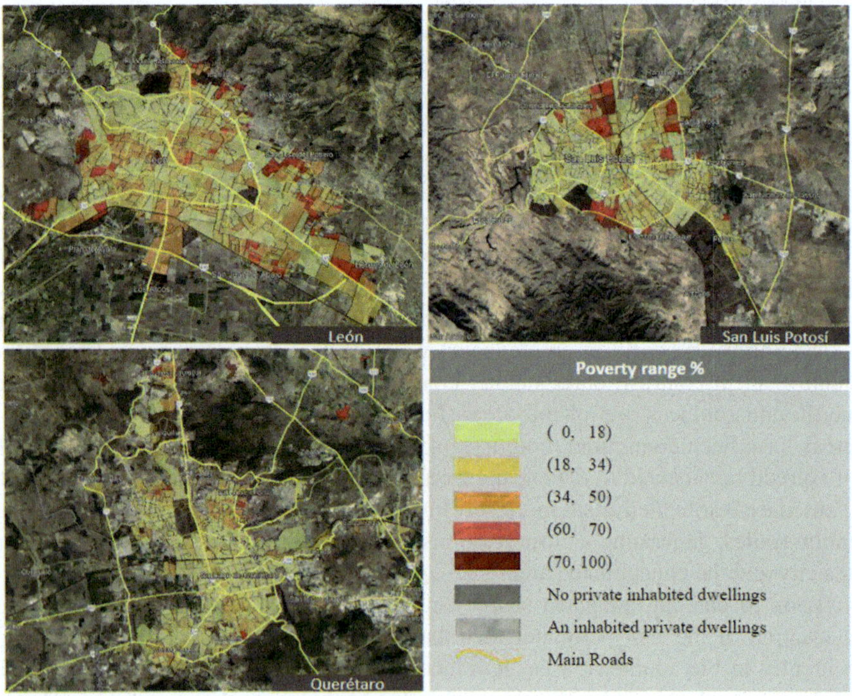


**Fig. 2.7** Ranges of urban poverty by BSGA in León, San Luis Potosí, and Querétaro in 2015. *Source* Google Earth, 2022

Among the three cities, León has the most significant number of BSGA with high poverty percentages. These are located more in peripheral areas to the northeast and west, within the containment perimeters destined for urban expansion. The consolidated areas present some contrasts: they contain BSGA with poverty rates above 34% and, in some cases, up to 70%, the latter to a lesser extent. Derived from this, it can be affirmed that León is a city with the highest disparities throughout its territory, despite being considered one of the best accessible, according to IMCO (2019).

By presenting a more delimited urban form (Fig. 2.7), in San Luis Potosí, the areas with the highest poverty rates are located in the north, northeast, and south of the city. They are contained in the second U2 and the third U3 perimeters. This shows that there are more homogeneous zones within the consolidated perimeters. Mostly, there is a population with low rates of registered poverty.

In the case of Querétaro, its territory's low poverty rates are presented, as there are few BSGA with rates above 60%. Among these, they are located in the expansion polygons. According to Salinas-Arreortua and Soto-Delgado (2019), the consolidated areas of Querétaro are not inhabited by those with lower economic resources. When contrasting these discoveries with IMCO data on inclusive cities (Fig. 2.5), Querétaro is not better positioned than the other two, especially when analyzing accessibility and the variables it implies.

## 2.5 Conclusions

The three study cities have similar population, climate, and topography characteristics. The shape of the routes in a city is determined by the connections it can have with others. In turn, these connections depend on the geographical location of the city. The main route is the backbone of the road system in a city. The most important is usually the route that gave rise to it and the road that connects one city with another. The main route for Querétaro and San Luis Potosí is Paseo de la República Avenue (or Federal Highway 57). For the city of León, Adolfo López Mateos Boulevard takes on the role of the main route.

The matrix routes that gave rise to the three cities have become the axes that have structured their growth. The elongated shape of Querétaro, from north to south on Paseo de la República Avenue, is the best evidence of the above. It is followed by San Luis Potosí. Although its most consolidated form is close to a circumference, for decades, an urban corridor has grown enormously from the entrance of the city (arriving from Querétaro) to the southeast, which reveals an elongated form on the Querétaro-San Luis Potosí highway (continuation of Federal Highway 57). Within the city of San Luis Potosí, it is also known as Paseo de la República Avenue.

A very different case is that of the city of León, whose shape is close to a circumference. This is due to its proximity to several cities, such as Guanajuato (to the southeast), Irapuato (to the south), San Luis Potosí (to the northeast), Aguascalientes (to the northwest), and Guadalajara (to the west). The shape of the city, structured by the roads that connect the city with those already mentioned, has guided growth along these routes.

Regarding the containment polygons, it was found that the "consolidated areas" are located, in the three study cities, along the main and some implantation paths, leaving the "areas without all services" around union paths. The less-favored population lives between the second and third containment polygon. This reinforces the current city production model, because worldwide, the only accessible land for low-income populations is the peri-urban. Peripherical sites do not have all the necessary services, and there are no critical roads (such as implantation routes or main routes), making it difficult for their inhabitants to move and exhibit, therefore, low accessibility.

Finally, it can be concluded that the form and structure presented by the cities of Querétaro, San Luis Potosí, and León favor the achievement of target 5 established as part of SDG 11, that is, to reduce the number of deaths caused by disasters, with particular emphasis in protecting the low-income population in situations of vulnerability. This is due to the topography of the three cities. The cases of dwellings located in risk areas due to steep slopes are minimal.

Cities today face significant challenges. The social, political, economic, and competitive dynamics they are subjected to have determined their course. The form is a consequence. The climate is an essential factor, but not limiting if energy savings must be favored. The form of the well-used and exploited territory, at the time of intervention in the city, can improve the inhabitants' condition and quality of life. Thinking of inclusive, safe, resilient, and sustainable cities is a provocation and, at the same time, a challenge that stems from political will and the strategic use of resources. Therefore, it is necessary to develop public policies that regulate urban growth and give guidelines for a better organization of the territory.

## References


Aguirre-Benítez K (2019) Manual de Calles. Diseño vial para ciudades mexicanas. SEDATU, Mexico City

Bazant J (1984) Manual de criterios de diseño urbano. Trillas, Mexico City

Campos-Sánchez F, Abarca-Álvarez F, Domingues A (2018) Sostenibilidad, planificación y desarrollo urbano: en busca de una integración crítica mediante el estudio de casos recientes. ACE 12(36):39–72

Caniggia G (1995) Tipología de la edificación. Estructura del espacio antrópico. Celeste, Madrid

Castro-Castro L (2018) Planeación sostenible de ciudades. Propuestas para el desarrollo de infraestructura. Fondo de Cultura Económica, Mexico City

Castro-Conrado Y, Fernández-Figueroa E, Álvarez-López A (2015) Influencia de la forma urbana en el microclima térmico de Sagua la Grande. Arq Urban 36(1):25–37

Data-México (2020) Ciudades y lugares. https://datamexico.org/es/explore?profile=geo&tab=2. Accessed 10 Jun 2022

Garza-Villarreal G (2002) Datos, hechos y lugares. Rev Inf Anál 19:7–16

González-Arellano S (2011) Ciudad desigual. Universidad Autónoma Metropolitana, Mexico City, Diferenciación socioresidencial en las ciudades mexicanas

Harvey D (1988) Social justice and the city. The Johns Hopkins University, Baltimore
Higueras E (1998) Urbanismo bioclimático. Criterios medioambientales en la ordenación de asentamientos. Instituto Juan Herrera, Madrid
Honnie SE, Grimm NB (2020) Nature-based approaches to managing climate change impacts in cities. Philos Trans R Soc B Biol Sci 307(1794):1–14
IMCO (2021) Centro de Investigación en Política Pública. https://imco.org.mx/indice-de-competitividad-urbana-2021/. Accessed 25 May 2022
INEGI (2022) Inventario Nacional de Vivienda 2020. https://www.inegi.org.mx/app/mapa/inv/. Accessed 12 Jun 2022
Lefevbre H (1968) El derecho a la ciudad. Península, Barcelona
Linares-García J, Hernández-Quirama A (2018) Accesibilidad espacial e inclusión social: experiencias de ciudades incluyentes en Europa y Latinoamérica. Civilizar 18(35):115–128
Naess P (2014) Urban form, sustainability and health: the case of greater Oslo. Eur Plan Stud 22(7):1524–1543
Ramírez B (2010) De la ciudad global a la ciudad neoliberal. Una propuesta teórica y política. In: Alfie M et al (eds) Sistema Mundial y nuevas geografías, 1a edn. Universidad Autónoma Metropolitana, Mexico City, pp 65–85
Rubalcava RM, Schteingart M (2012) Ciudades Divididas. El Colegio de México, Mexico City
Saldívar B, Badillo D (2021) Población mexicana crece a su menor tasa anual en 10 décadas: Censo 2020. https://www.eleconomista.com.mx/economia/Poblacion-mexicana-crece-a-su-menor-tasa-anual-en-10-decadas-Censo-2020-20210126-0001.html. Accessed 15 Jun 2022
Salinas-Arreortua L, Soto-Delgado L (2019) Política de vivienda en México: entre la expansión y el retorno al centro. Inv Geogr 99:1–17
SEDATU (2018) Sistema Urbano Nacional. Secretaría de Desarrollo Agrario, Territorial y Urbano. https://www.gob.mx/conapo/documentos/sistema-urbano-nacional-2018. Accessed 28 Jun 2022
SEMARNAT (2022) Atlas Digital Geográfico. Secretaría de Medio Ambiente y Recursos Naturales. http://gisviewer.semarnat.gob.mx/aplicaciones/Atlas2015/atm_climas.html#:~:text=En%20el%20territorio%20nacional%20se,Fr%C3%ADo%20con%20la%20m%C3%ADnima%20superficie. Accessed 18 May 2022
Sobrino J, Carrocho C, Graizbord B, Brambila C, Aguilar A (2015) Ciudades sostenibles en México: una propuesta conceptual operativa. CONAPO, Mexico City
Soja E (1996) Seeking spatial justice. University Minnesota Press, Minnesota
UN (2018) World urbanization prospects 2018. United Nations, Department of Economic and Social Affairs Population Dynamics, Population Division. https://population.un.org/wup/Download/. Accessed 25 May 2022
UN-Habitat (2020) Folleto de Datos Poblacionales 2020. Estado Global de las Metrópolis 2020. https://onuhabitat.org.mx/index.php/folleto-de-datos-poblacionales-2020. Accessed 17 May 2022
Vargas N, Bustos C, Ordoñez O, Calle M, Noblecilla M (2017) Uso y aprovechamiento de los recursos naturales y su incidencia en el desarrollo turístico local sostenible. Caso Pasaje. RIAT 13(2):206–221
Vidal-Zepeda R (2005) Las regiones climáticas de México. Instituto de Geografía-UNAM, Mexico City

# Chapter 3
# Construction of the Urban Sustainability Index for Cities in Mexico and Its Application in the Municipalities of Pachuca and Mineral De La Reforma, 2020

**Patricia Catalina Medina-Pérez, Jorge Villanueva-Solís, and Sergio Gabriel Ceballos-Pérez**

**Abstract** An urban sustainability index was constructed using the principal components method and census data disaggregated by basic statistical geographic area (BSGA) for the municipalities of Pachuca and Mineral de la Reforma in Hidalgo State, Mexico, in 2020. The variables were selected using a multiple regression method. The index allows knowing the degree of urban sustainability according to population density, services, and housing characteristics. As a result, it was obtained that 34% of the 364 BSGA are in the range of low and very low urban sustainability, 27% have moderate sustainability, and 38% present high and very high degrees of sustainability. In conclusion, this urban sustainability index could be used to design public policy criteria to improve the population's living conditions and replicated in any city in Mexico. However, it is necessary to incorporate new variables for future research.

**Keywords** Provision of urban services · Inclusive cities · Density · Sustainable Development Goals · Public policies

P. C. Medina-Pérez
Sustainable Urban Development, The College of the State of Hidalgo, Parque Científico y Tecnológico del Estado de Hidalgo Blvd. Circuito La Concepción, Exhacienda de La Concepción Número 3, 42162 Hgo, Mexico
e-mail: pmedina@elcolegiodehidalgo.edu.mx

J. Villanueva-Solís
School of Architecture, Autonomous University of Coahuila, Carr. Torreón - Matamoros KM. 7.5, Ejido El Águila. Ciudad Universitaria, C.P. 27276 Torreón, Coahuila, Mexico
e-mail: jorge.villanueva@uadec.edu.mx

S. G. Ceballos-Pérez (✉)
National Council for Humanities, Science and Technology (CONAHCYT) , Av. Insurgentes Sur 1582, Col Crédito Constructor, Alcaldía Benito Juárez, C.P. 03940 Ciudad de México, Mexico
e-mail: sgceballospe@conacyt.mx

 
L. Lizárraga-Mendiola et al. (eds.), *Sustainable Spaces in Arid and Semiarid Zones of Mexico*, The Latin American Studies Book Series,
https://doi.org/10.1007/978-3-031-53418-8_3

## 3.1 Introduction

The world has changed population growth and city concentration in the last fifty years. According to World Bank (2021), in 2020, almost 4.5 billion people (57% of the world's population) lived in cities, and according to projections, that number will increase to approximately 5 billion by 2030. This means that urban management and sustainability policies should focus on urban issues since cities concentrate most of the population and emit the most pollution (Kennedy et al. 2015).

In this sense, the Sustainable Development Goals (SDGs) targets 11.3 and 11.7 focus on sustainable urbanization and building sustainable, resilient, and inclusive cities and communities (UN 2022). Target 11.7 specifically aims to provide universal access to safe, inclusive, and accessible green and public spaces, particularly for women and children, older persons, and persons with disabilities. Target 11.3, on the other hand, aims to enhance the capacity for participatory, integrated, and sustainable human settlement planning and management in all countries. This target recognizes that sustainable urbanization requires the active participation of all stakeholders, including residents, civil society organizations, and local governments, in decision-making processes that shape the future of cities and communities.

The construction of indicators for urban sustainability contributes to cities' research, analysis, and planning; likewise, they give us an overview of the condition in which various aspects of urban sustainability are found (Gudmundsson and Regmi 2017; Rodrigues and Franco 2020). This chapter aims to build an indicator of urban sustainability through the principal components methodology and apply it to two semi-arid cities, Mineral de la Reforma and Pachuca in Hidalgo, Mexico.

## 3.2 A Conceptual Framework for Urban Sustainability and Indicators

### *3.2.1 Urban Sustainability*

There are various approaches to urban sustainability. Some emphasize the different environmental functions and the energy and material balance of cities; others assess environmental metabolism and socio-environmental conflicts (Rodrigues and Franco 2022), while other works consider in their conception inclusive social factors, such as access to decent housing, security, integration of vulnerable groups, and recreational or cultural spaces (Ceballos-Pérez 2017a; Shafie 2018). Other studies, for their part, include economic factors such as population density, development, specialization, and diversification of economic activities, which is reflected in greater efficiency and reduction of service costs, and access to employment, among others (Yan et al. 2018).

However, we must consider that any of these three approaches is valuable, and even a combination of them results in better development of the concept of urban sustainability since having most of these benefits is the goal of urban sustainability (Streule et al. 2020). Valcárcel-Aguiar et al. (2019) reaffirm this thought by breaking down four aspects of urban sustainability.

- The conservation of natural resources: this item aims to guarantee the stock of natural resources through the efficient use of the soil, the balance between ecosystem functions, the reduction of waste of non-renewable resources, and their replacement, whenever possible, with renewable resources and, finally, the maintenance of biological diversity.
- The built space: guarantees that the anthropic environment respects and is in harmony with the natural environment and that the relationship between the two is mutually productive.
- Environmental quality: consists of avoiding or reducing the processes that degrade or pollute the environment, protecting the ecosystem's regeneration capacity, and avoiding developments that harm human health or reduce the quality of life.
- Social equity: tries to avoid developments that contribute to widening the gap between the poor and the rich and support actions that reduce social inequities.
- In this way, urban sustainability is a strategy to move toward a better quality of life through changes in habits and responsible practices (Nikezić et al. 2021). Therefore, an approximation to urban sustainability would be defined as a space with a series of conditions for a high quality of life with a low ecological footprint, in which people act accordingly (Leclerc et al. 2019).

To define this more precisely, one can think of different issues of urban sustainability, such as mobility, water management, urban solid waste management, pollution management, land use planning, and conservation of ecological reserves and species, as well as many other topics that could be addressed (Brown et al. 1999; Stephen 2009; Ceballos-Pérez 2017b).

All these issues need to be addressed by specialists and with citizen, public and private sectors´ participation, to achieve substantive solutions (Leff 2002; Figueroa and Rotarou 2016). Accordingly, the viability of sustainable cities depends on various factors, in addition, to the technical aspects mentioned above social factors such as governments' political will, regulations, institutions, education, and communication, among others (Botello et al. 2014; Bueno-Amara and Fernandes-Cerveira-Filho 2016).

### *3.2.2 Urban Sustainability Indicators*

The design of indicators contributes to quantifying, registering, evaluating, comparing, and raising awareness about the urban sustainability of a specific space or region. Likewise, it contributes to the characterization of elements that we seek to know, such as population density, housing density, pollution indices, and proportion of green areas per person, among others (Cassiers and Thiry 2012; Benites and Simões 2021).

There are various types of urban indicators related to sustainability; some were developed from the New Urban Agenda (UN 2017), while others were designed to measure specific issues such as sustainable mobility, governance, citizen participation, multifunctionality of public space, green infrastructure, and urban biodiversity per $km^2$, among others (Sdoukopoulos et al. 2018; Michalina et al. 2021).

This chapter proposes the construction of an indicator of urban sustainability based on the provision of services and population density, which is the first step in which variables can be added according to the information available and the methodology of the main components.

## 3.3 Methodology

This work was integrated into two stages of analysis. The first stage presents the construction of the urban sustainability index (USI) through the principal components technique, with disaggregated census data from 364 basic statistical geographic areas (BSGA) for the year 2020 and the municipalities of Pachuca and Mineral de la Reforma, Hidalgo State, Mexico (Fig. 3.1). In the second stage the exploratory spatial data analysis techniques (ESDA), Moran's I and Local Indicators of Spatial Association (LISA), were applied. Finally, a spatial regression model was run to establish the relationships between these variables in the analyzed municipalities.

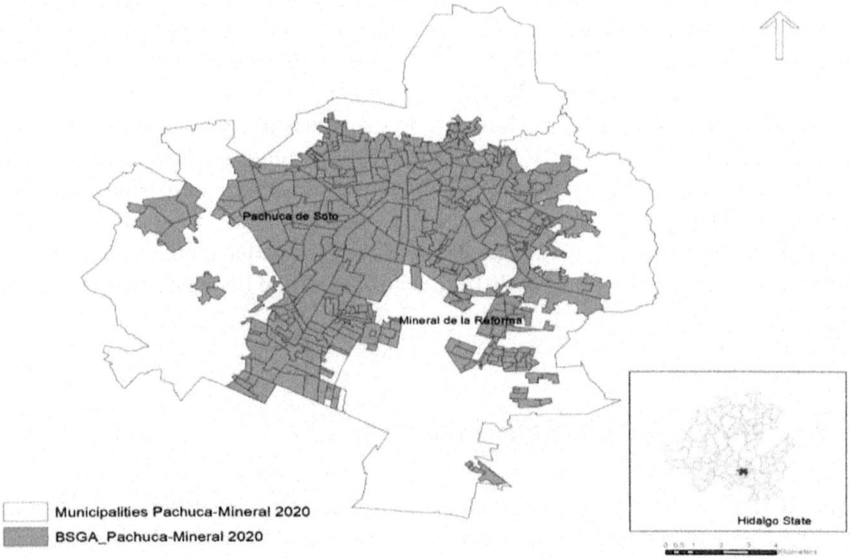


**Fig. 3.1** BSGA study zone of Pachuca and Mineral de la Reforma (Hidalgo State, Mexico)

**Table 3.1** Pachuca and Mineral de la Reforma descriptive statistics of the variables of the urban sustainability indicator at the BSGA 2020 level (N = 364)

| Variable | Indicator | Media | Standard deviation | N |
|---|---|---|---|---|
| %vph_celec | Private dwellings inhabited with electricity | 96.9568 | 13.80554 | 364 |
| %vph_aguadv | Private dwellings inhabited with water inside the dwelling | 94.2964 | 18.22404 | 364 |
| %vph_drenaj | Private dwellings inhabited with drainage | 97.0772 | 13.56993 | 364 |
| %vph_auto | Private dwellings with a car | 56.2900 | 20.55686 | 364 |
| %vph_refri | Inhabited private dwellings with refrigerator | 89.1130 | 16.20961 | 364 |
| %vph_lav | Private dwellings with washing machine | 74.3962 | 16.38319 | 364 |
| %vph_tv | Private dwellings inhabited with television | 92.1588 | 14.10565 | 364 |
| %vph_radio | Private dwellings inhabited with radio | 72.6770 | 12.80091 | 364 |
| %vph_pc | Private dwellings inhabited with a computer | 54.3597 | 21.15063 | 364 |
| %vph_telef | Inhabited private dwellings with telephone | 41.9469 | 21.20428 | 364 |
| %vph_cel | Private dwellings inhabited with a cell phone | 93.4410 | 13.43393 | 364 |
| %vph_inter | Private dwellings inhabited with the internet | 65.9269 | 20.13530 | 364 |

### 3.3.1 Urban Sustainability Index (USI)

The USI was built with variables related to access to essential urban services, the physical conditions of housing, and access to goods. The selection of variables has to do with the availability of related information; however, in the future work, it will be favorable to enrich the indicator with new variables that give a broader perspective of urban sustainability. The USI aims to measure the degree of sustainability and is built from the variables shown in Table 3.1, whose descriptive statistics of Pachuca and Mineral de la Reforma are presented, with information from INEGI (2020).

### 3.3.2 Exploratory Spatial Data Analysis (ESDA)

The ESDA allows showing spatial grouping patterns by identifying atypical values through statistics, maps, and graphs. Moran's I was used as an indicator of global spatial autocorrelation to test the hypothesis of the existence of data grouping patterns involved in a social phenomenon and whether these groupings are random or the result of spatial correlation (Odland 1988). Spatial autocorrelation exists whenever a variable exhibits a pattern over space in which its values at one set of locations depending on the importance of the same variable at other sites, which allows measuring the dependence between relative values in distribution from hypothesis tests on the distribution of the variables (Odland 1988).

Moran's I determines the degree of spatial association of the observations concerning the neighbors from a matrix of spatial weights as a neighborhood criterion. Moran's I values are between $-1$ and $+\ 1$; values close to $+\ 1$ indicate an association, while values close to -1 indicate dispersion. Moran's I can be understood in two terms: the data's variance and the regions' connectivity on a map as a single value, which measures the connectivity of such regions (Odland 1988).

The spatial conglomerates that Moran's I produces (Anselin 2018) can be classified into five categories: (1) low-low spatial units with values below the average, surrounded by units with values that are also lower than average; (2) low–high spatial units with values below average, surrounded by units with values above average; (3) high-low spatial units with values above average, surrounded by units with values below average; (4) high-high is formed by the spatial association of units with high values, surrounded by observations with high values; and finally (5) no data, where there is no correlation between the variable of interest and its neighbors (Ord and Getis 1995).

The techniques used to calculate spatial autocorrelation statistics can extend their use beyond testing the randomness of the patterns shown on the maps, allowing tests of alternative hypotheses about the spatial process that produces the patterns (Odland 1988). To the set of ESDA techniques, a focus on local patterns of association is added, in which two interpretations are combined: the evaluation of the significance of the groupings from the individual location and the indication of points of spatial non-stationarity or the suggestion of outliers or spatial regimes.

The local indicator of spatial association (LISA) is a statistic that meets two requirements (Anselin 1995): (1) the LISA of each observation indicates the extent of significant spatial clustering of similar values around that observation; (2) the sum of LISAs for all observations is proportional to a global indicator of spatial association.

The LISA decomposes the global autocorrelation index to know how much each spatial unit contributes to the general value, to simultaneously capture the degree of spatial association and the resulting heterogeneity of each contribution spatial unit (Celemin 2009). A significance value is obtained for each group formed by the similar values of each spatial unit and its neighbors; each location has a degree of uniqueness according to its site in space (Anselin 1995).

Spatial clusters (hot spots) are identified as groups of neighboring observations in which the LISA is significant. The LISA statistic can test the null hypothesis of no local spatial association (Anselin 1995).

### 3.3.3 Spatial Regression Models

If the ESDA makes evident the presence of spatial autocorrelation, it is necessary to run a regression model estimated with ordinary least squares to determine the influence of the explanatory variables on urban sustainability. Spatial regression models are based on the generalization of criteria based on the proximity between observations from a matrix of spatial weights (W). If the spatial dependence tests show that the autocorrelation persists, a spatial regression model must be selected because the autocorrelation and heterogeneity violate the homogeneity of variance and independence of the models' residuals by the ordinary least squares (OLS) method (Sánchez-Peña 2006). When spatial data is studied, the independence assumption yields inconsistencies due to spatial autocorrelation. When such dependence is found in the errors, they are called a spatial error model; when it is in the explanatory variables, they are called a cross-regression model (Anselin 2003).

The variables used in the construction of the regression model of the urban sustainability index are shown (Table 3.2), with information from the Mexican Population and Housing Census (INEGI 2020).

A Pearson correlation matrix was constructed to identify the presence of multicollinearity in the regression model; the variables observed in Table 3.3 were not significantly correlated, which is feasible for model construction. The results of the regression model in the Geoda program (Anselin 2002) confirm the existence of autocorrelation in terms of the error, which was corrected using a spatial model.

**Table 3.2** Descriptive statistics of the variables of the urban sustainability regression model at the BSGA 2020 level (N = 364)

| Variable | Media | Standard deviation | N |
|---|---|---|---|
| Percentage of population | 0.27473 | 0.207406 | 364 |
| Demographic dependency | 31.84624 | 7.560264 | 364 |
| Population aged 18 and over with post-basic education | 63.28176 | 20.939646 | 364 |
| Population in female-headed households | 31.57498 | 9.291989 | 364 |
| Private inhabited dwellings with different types of floors | 96.57626 | 13.871169 | 364 |
| One-bedroom private dwellings | 1.91841 | 3.351289 | 364 |
| Population density | 72.91910 | 53.270883 | 364 |

**Table. 3.3** Correlation matrix of the variables used in the spatial regression model

| Correlations | Depdem | Pob18ymaspb | Porcpob | Graproes | Hogjefem | Pisodt | 1cto | Denhab | Denpob Denpob | |
|---|---|---|---|---|---|---|---|---|---|---|
| Depdem | Correlation coefficient | 1 | −0.227** | 0.173** | −0.180** | 0.234** | −0.141** | 0.323** | −0.079 | −0.033 |
| | Sig. (bilateral) | | 0 | 0.001 | 0.001 | 0 | 0.007 | 0 | 0.136 | 0.526 |
| Pob18ymaspb | Correlation coefficient | −0.227** | 1 | −0.033 | 0.988** | −0.074 | 0.547** | −0.588** | 0.107* | 0.029 |
| | Sig. (bilateral) | 0 | | 0.528 | 0 | 0.162 | 0 | 0 | 0.042 | 0.578 |
| Porcpob | Correlation coefficient | 0.173** | −0.033 | 1 | −0.02 | 0.275** | −0.178** | 0.372** | 0.348** | 0.380** |
| | Sig. (bilateral) | 0.001 | 0.528 | | 0.704 | 0 | 0.001 | 0 | 0 | 0 |
| Graproes | Correlation coefficient | −0.180** | 0.988** | −0.02 | 1 | −0.052 | 0.516** | −0.560** | 0.08 | 0.006 |
| | Sig. (bilateral) | 0.001 | 0 | 0.704 | | 0.327 | 0 | 0 | 0.127 | 0.916 |
| Hogjefem | Correlation coefficient | 0.234** | −0.074 | 0.275** | −0.052 | 1 | −0.122* | 0.159** | 0.333** | 0.351** |
| | Sig. (bilateral) | 0 | 0.162 | 0 | 0.327 | | 0.02 | 0.002 | 0 | 0 |
| Vphpisodt | Correlation coefficient | −0.141** | 0.547** | −0.178** | 0.516** | −0.122* | 1 | −0.572** | 0.206** | 0.154** |
| | Sig. (bilateral) | 0.007 | 0 | 0.001 | 0 | 0.02 | | 0 | 0 | 0.003 |

(continued)

**Table. 3.3** (continued)

| Correlations | Depdem | Pob18ymaspb | Porcpob | Graproes | Hogjefem | Pisodt | 1cto | Denhab | Denpob Denpob | |
|---|---|---|---|---|---|---|---|---|---|---|
| Vph1cto | Correlation coefficient | 0.323** | −0.588** | 0.372** | −0.560** | 0.159** | −0.572** | 1 | −0.095 | −0.007 |
| | Sig. (bilateral) | 0 | 0 | 0 | 0 | 0.002 | 0 | | 0.071 | 0.894 |
| Denhab | Correlation coefficient | −0.079 | 0.107* | 0.348** | 0.08 | 0.333** | 0.206** | −0.095 | 1 | 0.988** |
| | Sig. (bilateral) | 0.136 | 0.042 | 0 | 0.127 | 0 | 0 | 0.071 | | 0 |
| Denpob | Correlation coefficient | −0.033 | 0.029 | 0.380** | 0.006 | 0.351** | 0.154** | −0.007 | 0.988** | 1 |
| | Sig. (bilateral) | 0.526 | 0.578 | 0 | 0.916 | 0 | 0.003 | 0.894 | 0 | |

**The correlation is significant at the 0.01 level (bilateral)
*The correlation is significant at the 0.05 level (bilateral)

## 3.4 Results and Discussion

Differentiated urbanization rates characterize Mexico's main cities according to the region in which they are located (Unikel 1976). The disparities are accentuated in those cities that do not have access to essential urban services to those with better living conditions. The popular and informal housing sectors show the most visible expression of urban poverty, characterized by limited access to land ownership, a built environment in poor conditions, vulnerability, overcrowding, and poor access to essential services (Prieto 2007).

Given that the construction of the urban sustainability index is based on the condition of accessible housing for urban services and goods, the areas that present high rates of urban sustainability are the regions of urban consolidation in the center of the city of Pachuca and the conurbation with Mineral de la Reforma. According to the analysis, 34.5% of the BSGA studied have low and very low urban sustainability (0–50). These areas are located mainly on the peri-urban and are characterized by low-income dwellings with low levels of income and scholarship. About 27.5% of the studied BSGA have moderate urban sustainability levels (50–75), while 38% belong to the high and very high sustainability (75–100).

The areas with low levels of urban sustainability are found in the northern, northwestern, northeastern peri-urban, and central sectors. The oldest sectors of the city present low levels of sustainability due to the deterioration of the infrastructure of public services; these are groupings characterized by popular dwellings that formed the first settlements of the city of Pachuca; actually, they remain inserted in the urban sprawl. The northern and northeastern peri-urbans are characterized by informality, are in popular settlements without provision of services, and are inhabited by migrants from the metropolitan area of the Valley of Mexico. Figure 3.2 shows groupings with low levels of urban sustainability. Using the ESDA technique made it possible to determine the location and significance of these groupings.

### *3.4.1 Global Spatial Association Indicator (Moran's I)*

Moran's I shows the existence of positive spatial autocorrelation of 0.668, with a significance level of $p < 0.05$. This explains the presence of BSGA groups with similar levels of urban sustainability in the urban areas of the principal municipalities of the Pachuca metropolitan area (Fig. 3.3). The spatial units considered were the 364 BSGA that make up the urban area of the municipalities of Pachuca and Mineral de la Reforma.

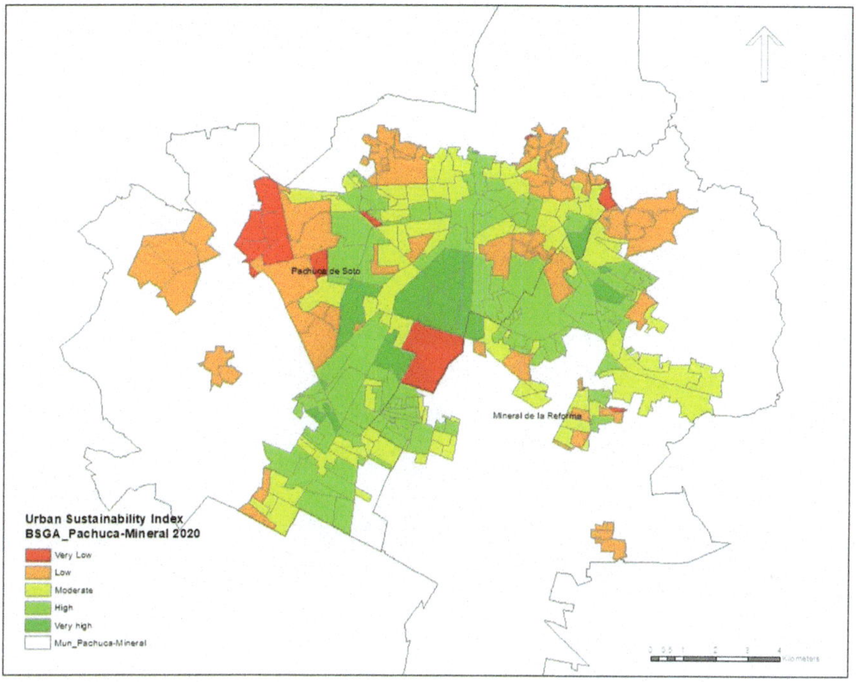


**Fig. 3.2** Pachuca-Mineral de la Reforma urban sustainability index, 2020

## *3.4.2 Local Indicators of Spatial Association (LISA)*

The LISA allowed identifying the local groupings around an individual location while providing the presence of spatial non-seasonality and suggesting the presence of atypical points in the study area. Figure 3.4 shows significant groupings in the high-high quadrant (intense red), with a positive spatial correlation; that is, there are BSGA with high urban sustainability values, surrounded by BSGA with high values. Two major high-concentration conglomerates of similar BSGA can be seen, located in the central and southern regions of the study area. Low-low clusters (intense blue color) also appeared, with a positive spatial correlation, which reveals BSGA with low levels of urban sustainability, surrounded by observations with the same characteristics. They are in the southeast and northeast sectors. Non-significant sectors in which no correlation between observations was found are shown in gray.

Figure 3.5 shows the statistical significance of the LISA, in which the groupings with high urban sustainability that are statistically significant are observed with darker green tones. Compared to Fig. 3.2, in which various sectors with high or low levels of urban sustainability could be observed, Fig. 3.5 shows significant and potentially valuable groupings for planning public policy interventions. The significant and highly significant clusters are found in the southeastern, central, and northeastern sectors of the municipalities of Pachuca and Mineral de la Reforma, with a statistical

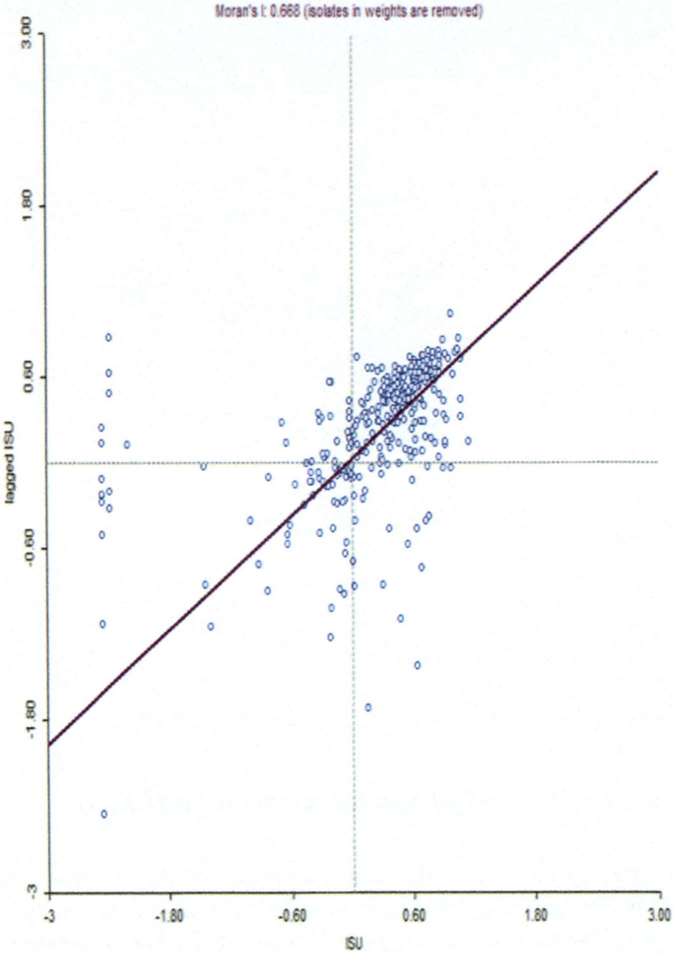


**Fig. 3.3** Moran's I of the urban sustainability index 2020 calculated for Pachuca and Mineral de la Reforma municipalities

significance of $p < 0.001$ for a total of 26 BSGA. The sectors with low levels of sustainability are found in the study areas' northwestern, northeastern, and southern peripheries.

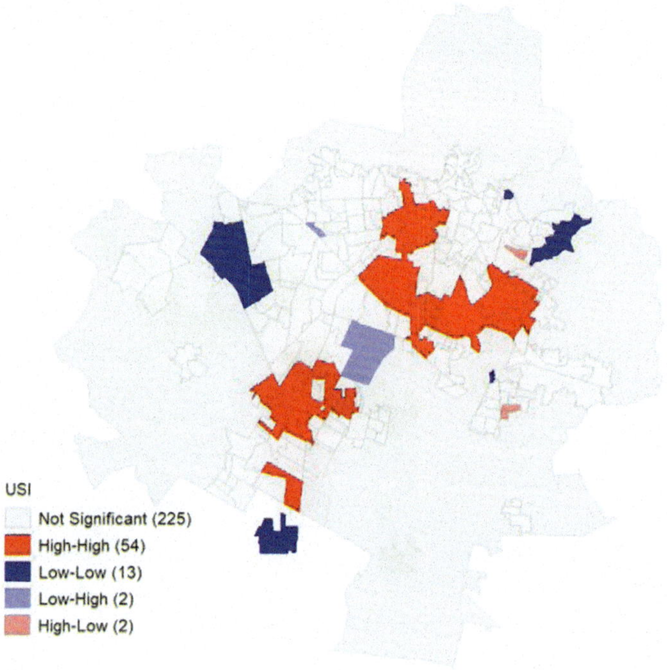


**Fig. 3.4** LISA of the urban sustainability index, 2020, calculated for Pachuca and Mineral de la Reforma municipalities

### *3.4.3 Identification of Variables that Explain Urban Sustainability (OLS Model)*

Once high and significant spatial autocorrelation in the variables studied with ESDA has been demonstrated, it is necessary to delve into the relationships between the variables studied and the USI. For the above, a multiple linear regression model was built using the ordinary least squares (OLS) method.

The fit value of the model estimated by OLS is $R^2 = 0.952096$, and the value of f-statistic (Fisher) $F = 1010.78$ and $p = 0.000$ provide a high statistical significance to the model. The variables that were significant in measuring urban sustainability according to the first model are shown in Table 3.4.

It is observed that the coefficient of the population percentage is positive and effective, which can be translated into a direct relationship between the amount of population with urban sustainability: the larger the population, the greater the sustainability. The same occurs with the variable population aged 18 and over with post-basic education, percentage of private dwellings inhabited with different types of floors, and demographic dependency. The rest of the variables were not statistically significant.

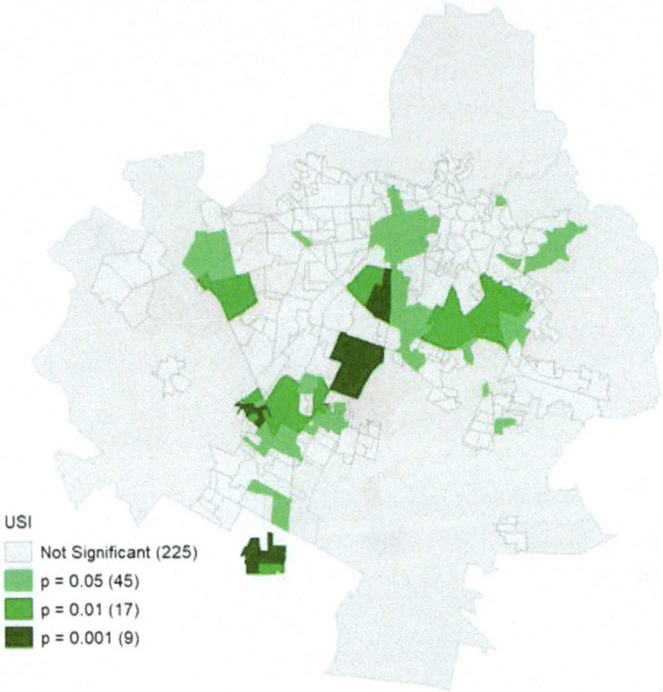


**Fig. 3.5** Statistical significance of the local indicators of spatial association

The regression model explains 95.2% of the USI variance, indicating that sufficient elements have been considered to explain urban sustainability in the study area. The diagnosis of spatial dependence suggests spatial autocorrelation (Moran's I 5.7742, $p = 0.000$), for which one of the principles of the OLS models are violated. So, a spatial regression model of the error was estimated, which has the highest value according to the value of the Lagrange multiplier (Anselin 2003).

In Table 3.4, the results of the spatial model of the error are also presented, for which a higher $R^2$ value was calculated in comparison with the version of the OLS model. The Akaike Information Criterion (AIC) in Table 3.4 measures the relative quality of a statistical model for a given dataset. The AIC provides a means for model selection; the smaller the AIC, the better the model fits the data (Anselin 2018). The AIC indicator is reduced in the error spatial error model from 2351 to 2322. The coefficients of the spatial model of the error did not change sign or probability, so the variables demographic dependency, the population aged 18 and over with post-basic education, the percentage of the people, and private dwellings inhabited with different types of floors continue to be significant to explain urban sustainability. By increasing the values of these variables, an increase in urban sustainability will be seen. In this way, to increase the living conditions of the population of the central municipalities of the Pachuca and Mineral de la Reforma urban area, efforts should be focused on

**Table 3.4** Ordinary least squares (OLS) regression model

| Variable | Estimated parameters | |
|---|---|---|
| | MCO model | Spatial regression model |
| Constant | 0.320169 | 0.551672 |
| Prob | 0.72258 | 0.61318 |
| Denpob | 0.0016881 | 0.256136 |
| Prob | 0.81119 | 0.000000 |
| Dep_dem | 0.290257 | 0.256136 |
| Prob | 0.000000 | 0.000000 |
| Hog_jef_f | -0.0238733 | 0.00225282 |
| Prob | 0.58226 | 0.95748 |
| Pob18ymas_pb | 0.545911 | 0.521815 |
| Prob | 0.000000 | 0.000000 |
| Porcpob | 5.33618 | 3.48264 |
| Prob | 0.00221 | 0.02999 |
| vph_1cto | 0.0105236 | 0.0116999 |
| Prob | 0.93582 | 0.92597 |
| Vph_pisodt | 0.297494 | 0.313255 |
| Prob | 0.0000000 | 0.000000 |
| Akaike criterion | 2383.06 | 2322.79 |
| R-squared | 0.952096 | 0.957212 |

the following dimensions: (1) physical conditions and access to urban services of the dwellings, (2) sociodemographic dimension, and (3) urban environment.

## 3.5 Conclusions

The results obtained allow, on the one hand, to demonstrate that it is feasible to develop an urban sustainability index through the principal components technique, in which public data from the national population and housing census is used. On the other hand, it identifies the degree of urban sustainability through its spatial features. This way, the proposed method was adequate to identify and analyze the degree of sustainability within the urban space.

Regarding the structure of the model, it showed its feasibility to be adapted to the urban-spatial conditions of any city in the country. If it is replicated in other cities or metropolitan areas, it would be possible to identify more significant sustainability relationships in the country's different regions, thus achieving a research framework adaptable to the national urban system. In this sense, the method and results presented here may be used to design public policy criteria to improve the population's living conditions, mainly concerning access to urban services, dwellings, and environmental quality.

Regarding the results, since the construction of the urban sustainability index is based on the access of the dwellings to urban services and goods, this relationship explains the existence of statistically significant spatial groupings, both at high levels and at low levels of urban sustainability, which is easily identifiable given the conditions of the urban context. Otherwise said, low levels of sustainability are related to areas of incipient urbanization and inadequate provision of services, such as occurs in the periphery of the study area and in the historic center areas, in which the infrastructure is obsolete and, in any case, the buildings intended for housing show significant deterioration. These observations are consistent with those found in other investigations in peri-urban areas and historic centers in Mexican cities.

This chapter presents a method that combines statistical analysis with geographic analysis to allow the study of the levels of urban sustainability in two suburban municipalities of central Mexico. Their spatial distributions and contrasting values generated spatial association schemes around sustainability. In this sense, the previous and following chapters link to sustainability through access to essential services and the preservation of natural heritage, respectively. Together they constitute a reference of analysis regarding the sustainability of human settlements. From what is described in this chapter and its links to the next one, the importance of access to essential and natural services as a point of reference for urban sustainability is evident.

## References

Anselin L (1995) Local Indicators of Spatial Association. Geogr Anal 27(2):94–115

Anselin L (2002) GeoDa software. The University of Chicago. https://geodacenter.github.io/ Accessed 13 May 2022

Anselin L (2003) Spatial regression with GeoDa. https://s4.ad.brown.edu/resources/tutorial/modul2/geoda3final.pdf. Accessed 13 May 2022

Anselin L (2018) An introduction to spatial data analysis. https://geodacenter.github.io/workbook/5a_global_auto/lab5a.html#morans-i. Accessed 13 May 2023

Benites AJ, Simões AF (2021) Assessing the urban sustainable development strategy: an application of a smart city services sustainability taxonomy. Ecol Indic 127:107734

Botello AV, Rendón-Von-Osten J, Gold-Bouchot G, Agraz-Hernández C (2014) Golfo de México Contaminación e Impacto ambiental. Diagnóstico y Tendencias. Universidad Autónoma de Campeche, Universidad Nacional Autónoma de México, Instituto Nacional de Ecología

Brown LR, Gardner G, Halweil B (1999) Beyond malthus: nineteen dimensions of the population challenge. Routledge Taylor and Francis Group, London

Bueno-Amara F, Fernandes-Cerveira-Filho JL (2016) Posibilidades de análisis de la acción socioambiental según la racionalidad ambiental en Enrique Leff. Tabula Rasa 24:171–189

Cassiers I, Thiry G (2012) Alternative indicators to Gdp: values behind numbers. Adjusted net savings in question. In: Sharpe A (compil) Applied Research in Quality of Life, spécial number

Ceballos-Pérez S (2017b) Landscape architecture and regional planning urban indicators in the metropolitan area of Pachuca. Hidalgo. Landsc Archit Reg Plan 2(1):1–12

Ceballos-Pérez S (2017a) Estudios de los problemas y realidad urbana actual en México. Colegio del Estado de Hidalgo

Celemin JP (2009) Autocorrelación espacial e indicadores locales de asociación espacial. Importancia, estructura y aplicación. Rev Univ Geogr 18:11–31

Figueroa EB, Rotarou ES (2016) Sustainable development or eco-collapse: lessons for tourism and development from Easter Island. Sustain 8(11):1093

Gudmundsson H, Regmi MB (2017) Developing the sustainable urban transport index. TD-ESCAP 87:35–53

INEGI (2020) Censo de Población y Vivienda, México. https://www.inegi.org.mx/programas/ccpv/2020/ Accessed 13 May 2023

Kennedy CA, Stewart I, Facchini A, Cersosimo I, Mele R, Chen B, Sahin AD (2015) Energy and material flows of megacities. Proc Nat Acad Sci USA 112(19):5985–5990

Leclerc A, Sala S, Secchi M, Laurent A (2019) Building national emission inventories of toxic pollutants in Europe. Environ Int 130:104785

Leff E (2002) Saber ambiental: sustentabilidad, racionalidad, complejidad, poder. Siglo XXI, México D.F. p 414

Michalina D, Mederly P, Diefenbacher H, Held B (2021) Sustainable urban development: a review of urban sustainability indicator frameworks. Sustainability 13(16):9348

Nikezić A, Ristić-Trajković J, Milovanović A (2021) Future housing identities: designing in line with the contemporary sustainable urban lifestyle. Buildings 11(1):18

Odland J (1988) Spatial Autocorrelation. Sage

Ord J, Getis A (1995) Local spatial autocorrelation statistics: distributional issues and an application. Geogr Anal 27(4):286–306

Prieto MB (2007) Condiciones habitacionales y calidad de vida urbana. El caso de la ciudad de Bahía Blanca. IX Jornadas Argentinas de Estudios de Población

Rodrigues M, Franco M (2020) Measuring the urban sustainable development in cities through a composite index: the case of Portugal. Sustain Dev 28(4):507–520

Rodrigues M, Franco M (2022) Bibliometric review about eco-cites and urban sustainable development: trend topics. Environ Dev Sustain 24:13683–13704

Sánchez-Peña L (2006) Métodos para el análisis espacial. Una aplicación al estudio de la geografía de la pobreza. II Congreso de la Asociación Latinoamericana de Población, Guadalajara, México

Sdoukopoulos A, Nikolaidou A, Pitsiava-Latinopoulou M, Papaioannou P (2018) Use of social media for assessing sustainable urban mobility indicators. Int J Sustain Dev 13(2):338–348

Shafie FA (2018) Barcelona towards sustainability. ajE-Bs 3(10):21–22

Stephen W (2009) Regions, megaregions, and sustainability. Reg Stud 43(6):63–876

Streule M, Karaman O, Sawyer L, Schmid C (2020) Popular urbanization: conceptualizing urbanization processes beyond informality. Int J Urban Region Res 44(4):652–672

UN (2017) The New Urban Agenda, The Conference on Housing and Sustainable Urban Development (Habitat III). United Nations

UN (2022) The sustainable development goals report. https://desapublications.un.org/publications/sustainable-development-goals-report-2022. Accessed 6 May 2023

Unikel L (1976) El Desarrollo Urbano de México: Diagnóstico e implicaciones futuras. El Colegio de México, México, D.F

Valcárcel-Aguiar B, Murias P, Rodríguez-González D (2019) Sustainable urban livability: a practical proposal based on a composite indicator. Sustainability 11(1):86

World Bank (2021) World urbanization prospects: 2018 revision. United Nations, Population Division

Yan Y, Wang C, Quan Y, Wu G, Zhao J (2018) Urban sustainable development efficiency towards the balance between nature and human well-being: connotation, measurement, and assessment. J Clean Prod 178:67–75

# Chapter 4
# Unearthing Disappeared Cultural Landscapes: The Case of the Old Tram Route of the Pachuca-Real Del Monte Mining District, Hidalgo, Mexico

**Carlos Alfredo Bigurra-Alzati, María Elena Sánchez-Roldán, Valeria Volpi-León, and Julio César García-Martínez**

**Abstract** The Pachuca-Real del Monte Mining District originated in the sixteenth century with the discovery of argentiferous veins. For their mining, it was necessary to draw paths, which became roads and corridors that were traveled by various types of transport. The topographical conditions and climate of the region influenced the definition of the routes that would later be traveled by railways and trams in the late 19th and early twentieth centuries. This research was carried out to recognize the historical stages of the routes in the mining district and the traces of railway tracks to promote the development of projects for tourist corridors that allow the revaluation of the landscape and its qualities, with a focus on the United Nations Sustainable Development Goal 11 –Sustainable Cities and Communities–. The route starts in the historic center of Pachuca and goes through the peri-urban area, which comprises the founding neighborhoods of this city, until reaching Real del Monte by rural roads. The traces of the routes that have been recorded and their current conditions were analyzed based on their soil type and potential for eco-tourist service stations, viewpoints, and supporting activities such as hiking and mountain biking, along with the tram rehabilitation. The goal is to contribute to the revaluation of the cultural

C. A. Bigurra-Alzati (✉) · M. E. Sánchez-Roldán · V. Volpi-León
Engineering and Architecture Academic Area, Autonomous University of Hidalgo State, Carr. Pachuca-Tulancingo km 4.5, Col. Carboneras. Mineral de la Reforma, C.P. 42184 Hidalgo, Mexico
e-mail: carlos_bigurra@uaeh.edu.mx

M. E. Sánchez-Roldán
e-mail: mariaesr@uaeh.edu.mx

V. Volpi-León
e-mail: volpi@uaeh.edu.mx

J. C. García-Martínez
Civil Engineering, National Technological Institute of Mexico, Felipe Angeles km. 84.5, Col. Venta Prieta, C.P. 42083 Pachuca de Soto, Hgo, Mexico
e-mail: julio.gm@pachuca.tecnm.mx

 
L. Lizárraga-Mendiola et al. (eds.), *Sustainable Spaces in Arid and Semiarid Zones of Mexico*, The Latin American Studies Book Series,
https://doi.org/10.1007/978-3-031-53418-8_4

landscape of the mining district by locals and visitors, as well as to contribute to the economic development and sustainability of the communities that make it up.

**Keywords** Ecosystem services · Landscape revaluation · Natural corridors · Tram routes · Urban–rural transition

## 4.1 Introduction

Pachuca de Soto is the capital city of Hidalgo State and is also the head of the Metropolitan Area of Pachuca, which groups seven adjoining municipalities: Pachuca de Soto, Mineral de la Reforma, Mineral del Monte, San Agustín Tlaxiaca, Epazoyucan, Zapotlán de Juárez, and Zempoala (Pérez-Torres 2013; Castillo-Ramos et al. 2015). Pachuca and Real del Monte developed together from a geographical location on the Neovolcanic axis (García-Arizaga and Lugo-Hubp 2003) in the Sierra de Querétaro and Hidalgo, currently called "Sierra de Pachuca," in Hidalgo State, Mexico (Fig. 4.1). It is known that its former inhabitants, notably Ñha-Ñhús, Chichimecas, and Nahuatl groups, were aware of the existence of various minerals in the territory. They mainly used obsidian to make tools and cutting instruments, such as axes and knives, and arrowheads as war weapons. To a lesser extent, they employed silver and gold in ornamental elements (Probert 2011).

In the sixteenth century, with the arrival of the Spaniards, the exploitation of silver and gold was prioritized (Chanfón-Olmos 1997). The communities with mining potential were called, for example, both "Mineral del Monte" and "Real del Monte" because during the viceroyalty, at least a fifth of the mining products had to be taxed

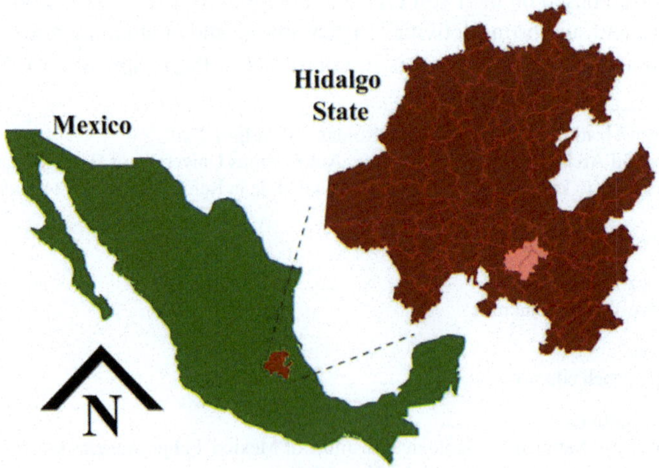


**Fig. 4.1** Study area (in pink) in Hidalgo State, Mexico

by the king of Spain. Nowadays, and in particular this municipality, it is known by both names (Probert 2011). Mining triggered the need to trace trails, which were first traveled on foot, then with the help of pack animals, and finally with carts and carriages dragged by mules and donkeys, which improved the transportation of workers, supplies, and minerals produced in the extraction process. However, the weather conditions, at an altitude between 2400 and 3000 m above sea level (masl), were problematic during autumn and winter. Also, in rainy seasons, the descent of water from the mountains dragged sludge and sediments produced by the mining work, which was recorded in the city's history due to the severe damage that the floods entailed (Lorenzo-Monterrubio 1995). These facts undoubtedly made the routes dangerous and encouraged the introduction of steam engines. In the nineteenth century, the first locomotives arrived (Wikiloc 2010; Ortega-Morel 2015); its use also represented the challenge of tracing the roads in the topography of the mountains, and the possible points for the route had to be analyzed considering the accesses to the mines. According to the reports issued by the Scientific Commission of Pachuca in 1864, there were about 80 registered mines between Pachuca and Real del Monte, so it was necessary to adapt roads to reach them, as well as to transport tools, inputs, workers, and the mining products (Comisión Científica de Pachuca 1865; Almaraz et al. 2014). The main areas that need to be communicated are currently occupied by the municipalities of Pachuca de Soto, Real del Monte, and Mineral de la Reforma (Fig. 4.2). Pachuca de Soto is an urban zone, passing through Mineral de la Reforma, which is part of the peri-urban, ending in Mineral de Monte, which belongs to the rural. For this reason, this transit includes now the complex transition and relationship from the urban and peri-urban to the rural contexts as part of the livability in outdoor spaces (Torres-Carral 2012; Galindo-Escamilla and Rodarte-García 2021; UN-Habitat 2021).

Given the importance of mining activities in the city toward 1880, Pachuca already had two railway stations, "El Central Mexicano" and "El Mexicano." Both companies transported mainly goods, not so many passengers. This caused the foundation of a new company, called "Ferrocarril Hidalgo y Nordeste", which occupied the land located between Belisario Domínguez and Gómez Pérez streets and between Juárez and Revolución avenues (Fig. 4.3). It should be noted that the paths to the mines were narrow because it crossed rugged topography in the mountains between Pachuca and Real del Monte. Hence, the dimensions of the regular railway cars were too large to travel them. For this reason, acquiring more compact model trams manufactured by the German company Siemens was essential (Morrison 2003).

In 1956, the offices and switchyard facilities were demolished to make way for the construction of the current Plaza Juárez in Pachuca; however, some buildings of this railway activity are still preserved, namely the building located on Calle de Mejía s/n, which from 1999 was called "Cultural Center of the Railway" and is nowadays headquarters of the annual Children and Youth's Book Fair. This contemporary reuse of the space also constitutes a testimony of the city's history for the knowledge of new generations.

Figures 4.4 and 4.5 show the traces of locomotives, railways, and trams that it was possible to recover from historical plans of Pachuca made between 1900 and 1926,

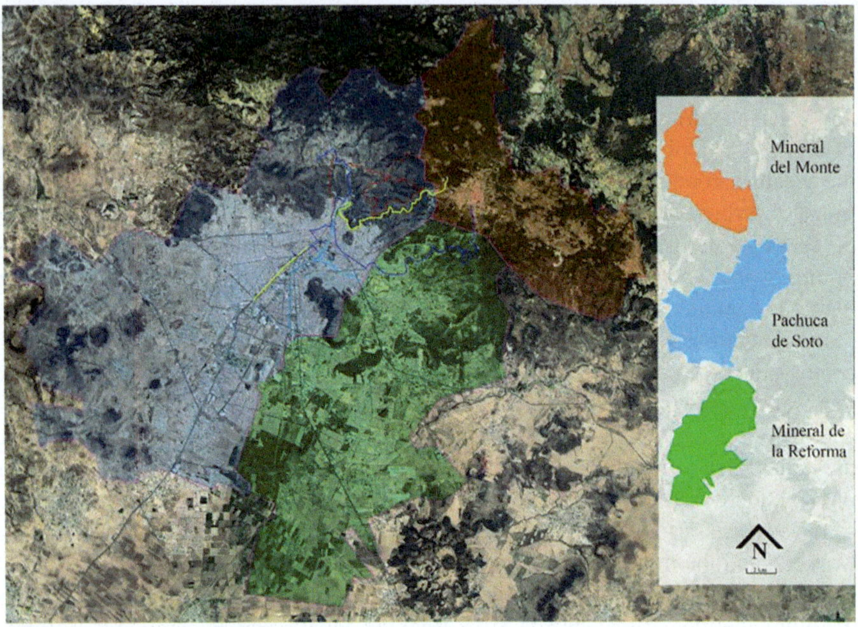


**Fig. 4.2** Study area and its three municipalities: Mineral del Monte, Pachuca de Soto, and Mineral de la Reforma. *Source* Google Earth, 2022

as well as from the sketch drawn by Morrison (2003). In both figures, besides the old train tracks (in blue), the current federal highway (in yellow) is also presented.

The present work aims to revalue the landscape and the historical context of the Pachuca—Real del Monte mining district by identifying and analyzing the registered old road traces and their new conditions. After the revision of the climatic and physiographic characteristics of the zone, soil samplings were performed to get information about the soil types, the strata that compose them, and their carrying capacity to unearth the route and denote future hiking, biking, eco-tourism stations, mountain biking, and tramcar projects, which may promote the social, cultural, and economic development of the communities that comprise it. In this way, we aim to contribute to the Sustainable Development Goal (SDG) 11 –Sustainable Cities and Communities– (UN 2022).

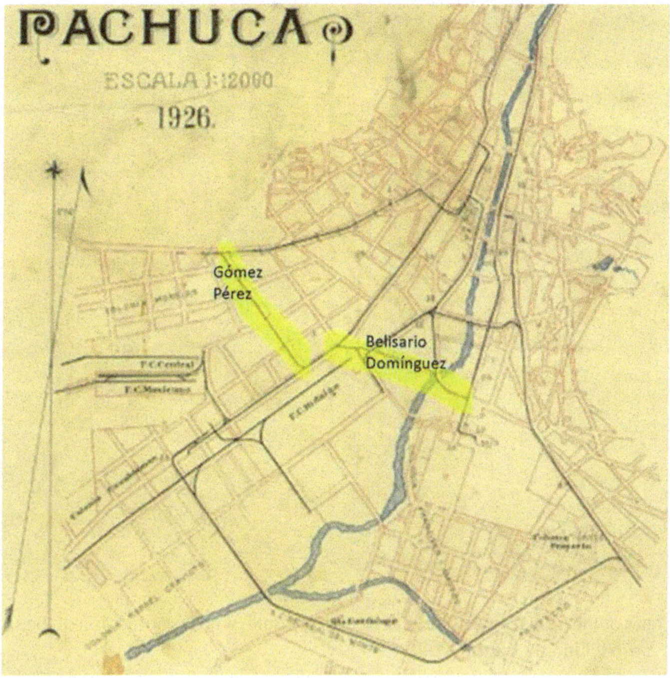


**Fig. 4.3** Departure of the train, tram, and roads of the center of Pachuca (in yellow). Location plan of 1926 with code registration 1926-CGE-7246-A, found in the Mapoteca Manuel Orozco y Berra. *Source* SCEH (2017)

## 4.2 Methodology

### *4.2.1 Climate of the Study Area*

According to the scientific nomenclature, the locality's general climate is a semi-arid zone. The climate was addressed considering two sources (Table 4.1); the climatic data obtained from the meteorological stations of CONAGUA (n.d.) and the climate map prepared by the National Institute of Geography and Statistics (INEGI n.d.a; García-Amaro 2004), where three types of climates in the study area were identified (Fig. 4.5). These climates prevail in the locations described below.

- Subhumid, tempered climate (in green in Fig. 4.5). This climate prevails in much of the municipality and the town of Mineral del Monte, including the center, north, east, west, and part of the south (INEGI n.d.a; INEGI 2005).
- Subhumid semi-cold tempered climate (in blue-green in Fig. 4.5). It includes part of the municipalities of Mineral de la Reforma and Pachuca (López-Levi and Valverde-Valverde 2015). This climate only prevails in a small south-southeast part of Real del Monte municipality.

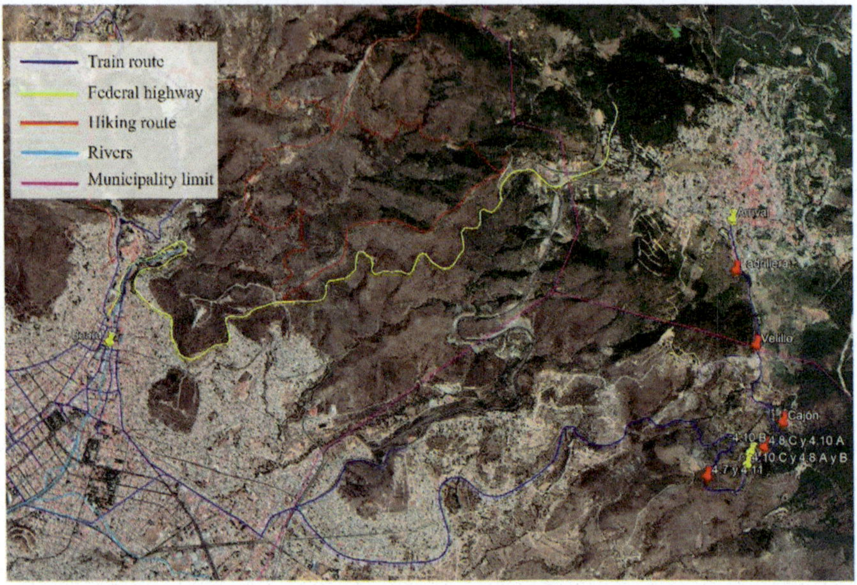


**Fig. 4.4** Route of the train from Pachuca to Real del Monte, with identification of the soil samplings carried out. *Source* Google Earth, 2022

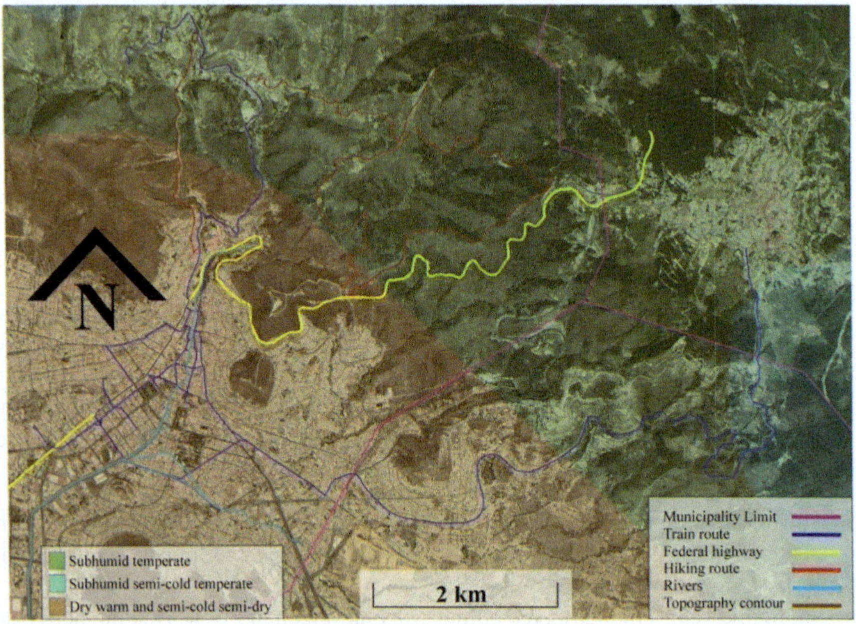


**Fig. 4.5** Climates of the study area. *Sources* INEGI (2017) and Google Earth, 2022

**Table 4.1** Description of climatic data obtained from the meteorological stations of CONAGUA, its nomenclature, and its description

| Weather station[a] | Location | Type of climate | Description[b] |
|---|---|---|---|
| CONAGUA—13,150 | San Miguel Cerezo | C(w1)bigw" | Temperate; the average annual temperature is between 12 and 18 °C. The coldest month is between −3 and 18 °C. The temperature of the hottest month is 6.5 - 22 °C; isothermal oscillation < 5 °C. An intermediate degree of humidity between C (w0) and C (w2), and rainfalls in long cool summers |
| CONAGUA—13,115 | Real del Monte | C(w2)bigw"/ updated C(w1)(w)bigw" | Temperate; the average annual temperature is between 12 and 18 °C. The coldest month is between −3 and 18 °C. The average temperature of the hottest month is 6.5–22 °C; isothermal oscillation < 5 °C. Intermediate humidity degree between C(w0) and C(w2), with rainfalls in long cool summers. P/T ratio between 43.2 and 55.0; percentage of winter rainfall < 5 of the annual |
| CONAGUA—13,022 | Pachuca de Soto | BS1kwigw"/ upgraded to BS1wk'igw" | Temperate with a cool summer; the yearly average temperature is between 12 and 18 °C; the coldest month is between −3 and 18 °C and the hottest month > 18 °C; isothermal oscillation < 5 °C. The percentage of winter rainfall is between 5 and 10.2 of the annual total |

[a]CONAGUA (n.d.)
[b]Pavón-Hernández and Meza-Sánchez (2009)

- Dry-warm and semi-cold semi-dry climate (in dark and light brown in Fig. 4.5). It includes part of the municipalities of Pachuca and Mineral de la Reforma. This climate prevails in the center, south, and most of the east and west parts of the map.

### *4.2.2 Physiography of the Study Area*

Geographically the study area is located within the Neovolcanic belt, characterized by a vast mass of volcanic rocks of all types, in the center of Mexico (Krasilnikov et al. 2011, 2013). The old tram route begins in Pachuca city, with a latitude of 20°05′15" N, a longitude of 98°44′59" W, and an altitude of 2766.0 masl. It ends in Mineral del Monte, with a latitude of 20°07′59" N, a longitude of 98°40′09" W, and an altitude of 2766.0 masl.

Mineral del Monte is between the plains and sierras of Querétaro and Hidalgo (García-Arizaga and Lugo-Hubp 2003; Krasilnikov et al. 2013). It extends west of Querétaro city to Pachuca, with 7821.33 km$^2$ within Hidalgo state. It represents a corridor, below 2000 masl, of low hills of volcanic material. It is practically enclosed from all sides by systems of mountains, plateaus, and hills, almost all of volcanic origin, within the subprovinces Lakes and Volcanoes of the Anahuac, where the study area is located. The Regosol type of soil is abundant and shows similar layers of light tone on mountain slopes (INEGI n.d.a). These are stony soils of light tones; they are similar to the rock that has given rise to them when they are not deep. They are common in the mountainous or arid regions of Mexico (INEGI 2010; INEGI 2017). It is associated with the Phaeozem soil type in a smaller area, which is prominent in semi-dry and subhumid climates. It has a brown-to-black surface color, is rich in magnesium, potassium, and lacks of carbonates in the subsoil. The relief where these soils develop is generally flat or slightly mountainous. They are essential soils for agriculture (INEGI n.d.b).

In Fig. 4.6, it is shown that the surface geological classification of the study area is ToA-BvA (andesite–breccia–andesite), which are formations defined by Geyne et al. (1963) as the cortex. Pachuca and Real del Monte present massive andesitic spills with gradational to dacitic changes; this unit predominates by far throughout the northeastern vertex of the geological-mining chart, and so constitutes the vast majority of the mineral deposits of the region.

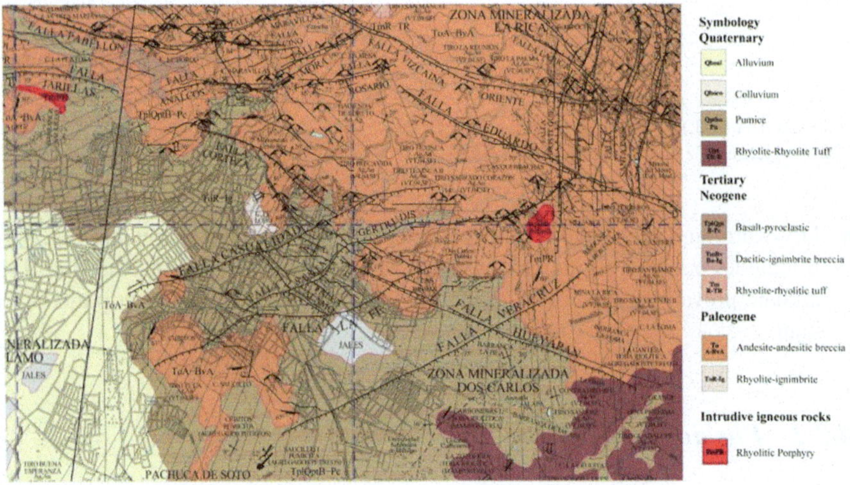


**Fig. 4.6** Geological classification of the study area. *Source* SGM (1997)

### *4.2.3 Analysis of the Historical Tram Route*

Land use (INEGI n.d.c), climate (INEGI n.d.a), and altitude, among other factors, determine the vegetation type. In the study zone (Table 4.1), the climate ranges from the semi-dry temperate of Pachuca de Soto and Mineral de la Reforma (with the use of land and vegetation of rainy season forest and arid zone scrub) to the semi-humid (temperature and semi-cold) from Mineral del Monte, with land use and vegetation of coniferous, oaks, and rainy season forest. Due to the above, we notice the relevance between the urban, peri-urban, and rural relationships, in which climate and vegetation are blended subtly into the landscape (Fig. 4.7).

Currently, few vestiges remain as tangible evidence of the route of the trams destined for mining work between the municipalities of Mineral del Monte and Pachuca, Hidalgo (Fig. 4.8). In the twentieth century, following the decline in mining production and several labor problems, the Mining Company removed all the infrastructure of the railways. The memory of the historical route is preserved by scarce documentary evidence, which is included in various sections of roads on maps made by authors such as Almaraz et al. (2014) and in the oral tradition shared through generations (Fig. 4.8).

**Fig. 4.7** Landscape view in a segment of the historical route

**Fig. 4.8** Historical tram route: **a** Original tram; **b** and **c** current views of the tram

To verify the existence of the route followed by the tram, it was necessary to carry out excavations in different sections to analyze the soil and corroborate if its capacity was sufficient to support a structure corresponding to cargo transport all throughout the study zone. It was also fundamental to identify the type of drainage work that could be put in place to verify its functionality for evicting the mine runoff and supporting the transport infrastructure.

The starting point for the soil samplings was on La Estación street, located in the Guadalupe neighborhood, in the municipality of Mineral del Monte, Hidalgo (Fig. 4.4). The final soil sampling was carried out at the point called "Mina Dos Carlos," currently located in the San Guillermo La Reforma neighborhood, in Mineral de la Reforma, Hidalgo (Fig. 4.4).

### *4.2.4 Soil Analysis of the Historical Path*

Representative soil samples were obtained in different depths, by hand or with a sampler, in some spots of the study area (Fig. 4.4) to get information about the soil types, the strata that compose them, and their carrying capacity (SICT 2003). Once the material samples were obtained, laboratory tests were carried out to determine water contents, liquid limits, plasticity indices (ASTM 2020a), and granulometry to classify the soil by the Unified System for Soil Classification (ASTM 2021a). The material conformation was determined through visual observation (ASTM 2021b); it includes the particle size, color, and the content of gravel and sand. The soil samples were tested in the triaxial chamber to determine their load capacity (ASTM 1996). Also, cohesion, internal friction angle, and density were measured (ASTM 2020a, b; ASTM 2021c).

## 4.3 Results and Discussion

### *4.3.1 Geotechnical Characteristics of the Historical Tram Route*

First, we found that the maximum slope of the historical tram route point was + 4.09%, and the minimum point was 1%, with an average gradient of 3.75% (Fig. 4.9). The track width of the cross section was 4.2 m and 3.54 m at the widest and narrowest points, respectively. We found several soil types, from clayey with high plasticity, combined with sands to andesite rock, in which the load capacity was between 12.12 tons/m$^2$ and 225.6 tons/m$^2$. These results, summarized in Table 4.2, indicate that diverse activities, such as biking, hiking, recreation, and even the rehabilitation of the tram infrastructure, could be carried out on the historical route.

As shown in Table 4.2, four soil samples were obtained along the tram route, the first of which was taken from a rural area.

The second sampling was performed in the “El Velillo” location, where a masonry “vault” type drainage work (DW1), with a length of 15.40 m, 1.40 m height, and 1 m width, was found. It is presumed that it is a work of the nineteenth century, corresponding to the tram execution period (SOP 1965; SCT 2021). Following the route, a second drainage work (DW2) was identified, similar to DW1. DW2 also has a masonry “vault” type, with a length of 19 m, 1 m height, 1.05 m width, 0.52 m radius, a distance of 2.5 m from the foundation, plus a crown with a 3 m width.

It was possible to make out the fourth soil sampling thanks to the neighbors’ collaboration in the area because it is currently located on private property. It was one of the most representative explorations of the study site because it has remained unchanged since the mining company removed all the railway infrastructure from the area (Fig. 4.10b). Figure 4.10c shows a drawer-like section with a slope of 1/4 in both cuts, where the caisson developed a depth of 10 m.

No samplings were carried out from Km 4 + 250 to Km 7 + 238 in the section corresponding to the route’s endpoint because the road leads to private properties. Also, from Km 7 + 238 to Km 7 + 877, there were no more explorations because it is currently in an urbanized area, and there is additional induced work, such as drainage, sewage, drinking water, fiber optics, and telecommunication lines.

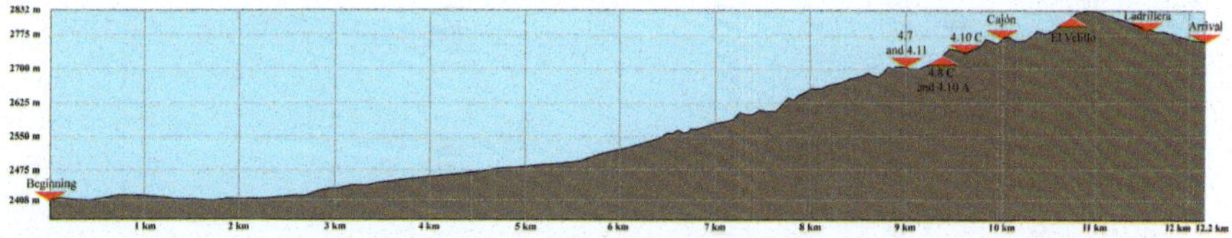


**Fig. 4.9** Cross section of the tram route, from the beginning to the arrival points, also signaled in Fig. 4.4

**Table 4.2** Description of soil samples taken on the tram route

| Sample | Location | Coordinates (m) | | Km | Site description | Cross-section | Slope (%) | Soil characteristics | Load capacity |
|---|---|---|---|---|---|---|---|---|---|
| | | **E** | **N** | | | (m) | | | (ton/m$^2$) |
| 1 | La Ladrillera | 533,916.17 | 2,225,991.09 | 0 + 440 | – | 3 | 4.09 | Clay with high plasticity sands and orange clay sands | 12.12 |
| 2 | El Velillo | 534,021.45 | 2,225,318.53 | 1 + 214 | – | 4 | 3.79 | Layer of gray rock with white dots | 225.6 |
| 3 | El Velillo | 534,214.00 | 2,224,645.00 | 2 + 980 | Balcony section with a slope of 1/4 | 3.5 | 3.54 | Mountainous terrain. Geological formation of andesite rock—gap—andesitic with many regional mineral deposits. Light gray rock with white dots | 173.3 |
| 4 | El Velillo | 533,864.47 | 2,224,299.21 | 3 + 574 | Drawer-like section, with a slope of 1/4 in both cuts | 4.2 | 1 | Clays and sands were identified in the first soil stratum, and two types of rocks in the second: lilac graneodite with white dots and pink limestone with white dots. The third stratum had cream-colored quarry rock | 22.11 |

Fig. 4.10 Landscapes of the old tram route are now hiking routes

A rail section was recovered at Km 3 + 048, where the second drainage work was also found (Fig. 4.11). The element was analyzed according to European and American regulations. The measurements obtained from the rail were 105 mm of standard bottom length rail, 105 mm height, 12 mm soul width, and 54 mm top length (Fig. 4.12). The element's weight was 29.3 kg. When comparing the piece with the European standard (Aceros CP n.d.), it was determined that the rail section obtained in the study area coincides with the S-30 type rail model, whose dimensions are 108 mm in bottom length, 108 mm in height, 12.3 mm of soul width, and 60.3 mm of top length (Fig. 4.12). There is a slight variation in the piece's dimensions with the original due to wear and erosion over time. Figure 4.12 shows that German tramcars crossed the narrowest areas of the route because the wagon was slim and light, which was possible due to the comparison between the European standards Deutsche Industrie Norm (Aceros CP n.d.) and the measurements of the rail found on one of the narrowest roads.

Concerning DW1 and DW2, they were built with rocks from nearby material banks because they have high resistance and were joined with mortar thicknesses of

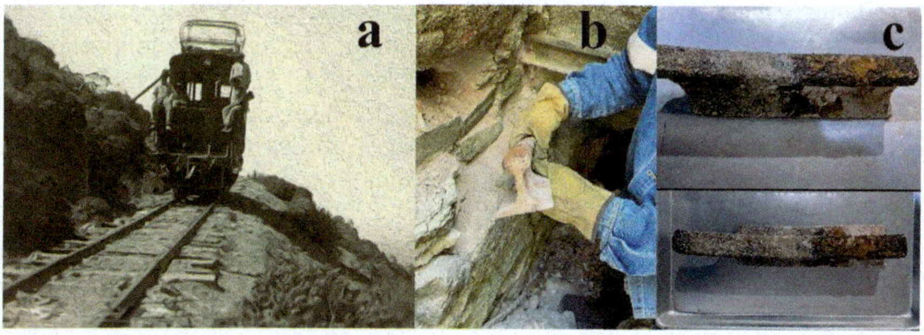

Fig. 4.11 Rail model of the track used for the tram

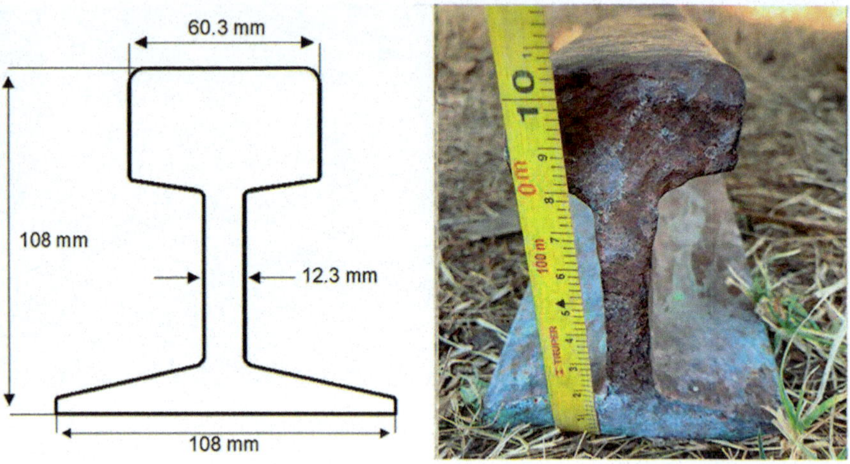


**Fig. 4.12** Type of rail used in the tram

around 3–4.5 cm. Their simple compressive strength was 98.66 kg/cm$^2$. Following the vestiges of drainage works for roads (SOP 1965), it was observed that the vaults were built with the same old design. This could correspond to endeavors to drain the mining sludges and avoid floods on the tram route (Almaraz et al. 2014).

Overall, we conclude that the physical characterization of the soil, the drainage works, and the analysis of the rail pieces obtained from the area confirm the specific type and model of the tram and the historical route, which is consistent with the documentary evidence.

### *4.3.2 Revaluation of the Mining Tram Route*

To explain the process of revaluation of a resource, it is important to start with the concept of value, whose definition was taken from Pontón (1972). The term comes from Latin (value, ~ õris) and refers to the degree of utility or aptitude of things as the "quality of things that are worth, that is, that they have a certain merit, utility or price." It, therefore, obliges a responsibility for its protection and safeguarding. The process of valuation of objects, movable and immovable, has been long and complex because to give value to something, it is necessary to make a critical, thorough, and reflective judgment that allows to issue or not, a value judgment that can be considered as a "cultural asset."

According to Choay (2019), heritage should be understood as a legacy from our ancestors. Prior generations must transfer it to us to be known, enjoyed, and valued. However, the author warns that the value given to the elements, whether monuments, buildings, places, or objects, will depend on the judgment and care that each generation confers on them. The foregoing is especially relevant today since

SDG target 11.4 seeks "to protect and safeguard the world's cultural and natural heritage" (UN 2022).

To define the valuation criteria to designate a monument, it is essential to quote Riegl (2017), an Austrian historian, who considered two aspects: (a) the commemorative value, which includes the evolution of the monument, its antiquity, and history, as well as intentional remembrance, and (b) the contemporary value, which includes the instrumental and artistic value. This classification responds to the fact that Riegl (2017) considered that different treatments are required for the conservation and/or restoration of the element to maintain in the people's conscience the experiences, feats, or historical facts worthy of value.

The mining activity that promoted the use of the tramways and their respective routes between the mountains should be revealed, mainly to disseminate the historical value of this infrastructure and to procure its intentional remembrance by the contemporary population. This disappeared infrastructure is an important piece in the territory's history and persists in the memory of the ancient generations still present in the zone.

The old tram route identification and legitimacy were supported by bibliographic evidence of maps showing part of the route. The process was corroborated at the beginning through observation of the area. In each section examined, it was noticed that the width of the railway was always greater than 3 m to embrace the dimensions of a light tram car. The slopes never exceeded 4.1%, avoiding forcing the machinery of the trams. The analysis of samples showed a load capacity between 12 and 225 tons/$m^2$ (Table 4.2), which is enough to support the weight of a light tram weighing 3.5 tons per wagon. Likewise, the piece of rail recovered is compatible with European rail models operating at the time, which has to be confirmed by proper historical research.

As part of the tram route assessments in what today is given as cultural walks with the appreciation of the landscape, the climate becomes essential due to the diversity of flora species. In these tours, the coexistence of the temperate semi-dry to the semi-cold subhumid climates was appreciated (Table 4.1), as well as the variation from xerophytic vegetation to coniferous forest.

Due to the above, safe routes and walks are recommended in the year's dry season, where the months with less rainfall are November, December, January, and February. On the other hand, dangerous routes due to wet and water-saturated soils are likely to occur in the rainy months (from June to September).

The passage of time has dusted the memories of old inhabitants and users, which forces people to revalue exhibited elements to obtain the population's recognition of historical stages, and the interest to know, revalue, rescue, and preserve them. So, it is conceivable to carry out projects evoking fragments of the mining tradition in the region through ancient transportation. The above might strengthen cultural identity by recovering traditional routes through promoting the social, economic, and cultural development of the communities that make up the mining district.

## 4.4 Conclusions

Field research made it possible to corroborate the routes and traces of the old railway and tram tracks. The guidance of current inhabitants of the communities allowed us to find material evidence of railways. This fact enables new stakeholders to learn fragments of the city's history to reinforce the identity of this territory and promote positive economic, social, and environmental links in urban and peri-urban areas. Likewise, it was evidenced that endemic vegetation (xerophilous foliage) currently covers the study area, which contributes to an appealing landscape with great interest for cycling and hiking users. The rehabilitation of the tram route pretends to promote the interest in learning about the local history, stimulate future proposals that improve eco-tourism services, and safeguard users. However, it is essential to note that the studied area is recognized only by the customs and habits of a few people because there is no proper signaling and promotion, and it is at their own risk to take over the historical route.

The legacy of the mining towns is full of valuable information from various angles. On the one hand, using local construction materials and eco-building processes is recommended to continue with the typology of nearby towns regarding natural corridors and urbanization. On the other hand, we underline the prevailing need for the protection of local ecosystems and the development of public policies that include the social and economic growth of the peri-urban communities. This research has shown technical, social, and climatic aspects, which will be necessary to deepen. The dialogues and efforts between those who inhabit the area, those who are interested in knowing about it, local authorities, researchers, students, and society, would allow keep alive the cultural heritage of the Mexico's mining region.

## References

Aceros CP (n.d.) Riel liviano, estándar europeo, norma DIN 5901. http://www.aceroscpsac.com/producto/riel-liviano-estandar-europeo. Accessed 8 Sep 2022

Almaraz R, Rivera-Cambas M, García-Cubas A, Meneses-Llaguno JM (2014) Memoria de los trabajos ejecutados por la Comisión Científica de Pachuca en el año de 1864. Miguel Ángel Porrúa - Gobierno del Estado de Hidalgo, Pachuca de Soto, Mexico, pp 285–354

ASTM (1996) ASTM D3999–91. Standard test methods for the determination of the modulus and damping properties of soils using the cyclic triaxial apparatus. Am Soc Test Mater

ASTM (2020a) ASTM-D4318–05. Standard test methods for liquid limit, plastic limit, and plastic index of soils. Am Soc Test Mater

ASTM (2020b) ASTM D5080-08. Standard test method for rapid determination of percent compaction. Am Soc Test Mater

ASTM (2021a) ASTM D2487. Standard practice for classification of soils for engineering purposes (Unified Soil Classification System). Am Soc Test Mater

ASTM (2021b) ASTM D2488-06. Standard practice for description and identification of soils (Visual-Manual Procedure). Am Soc Test Mater

ASTM (2021c) ASTM D5195–21. Standard test method for density of soil and rock in-place at depths below surface by nuclear methods. Am Soc Test Mater

Castillo-Ramos I, Ornelas-Delgado J, Hernández-Cortés C (2015) Las zonas metropolitanas, reflexiones teóricas y estudios en el centro del país. Universidad Autónoma de Tlaxcala - Miguel Ángel Porrúa, Mexico City, pp 17–44
Chanfón-Olmos C (1997) Historia de la arquitectura y el urbanismo mexicanos, Volumen II - El periodo virreinal. Tomo I - El encuentro de dos universos culturales. Fondo de Cultura Económica - UNAM, Mexico City, pp 189
Choay F (2019) Alegoría del patrimonio. Gustavo Gili, Barcelona
Comisión Científica de Pachuca (1865) Memoria de los trabajos ejecutados por la Comisión Científica de Pachuca en el año de 1864. https://memoricamexico.gob.mx/es/memorica/Comision_Cientifica_de_Pachuca. Accessed 8 Sep 2022
CONAGUA (n.d.) Normales climatológicas por estado. Estado de Hidalgo. Comisión Nacional del Agua. https://smn.conagua.gob.mx/es/climatologia/informacion-climatologica/normales-climatologicas-por-estado?estado=hgo. Accessed 8 Sep 2022
Galindo-Escamilla E, Rodarte-García R (2021) Pachuca y su relación campo—ciudad: Tierra, agua y agricultura en los siglos XX y XXI. Universidad Autónoma del Estado de Hidalgo, Pachuca de Soto, Mexico, pp 15–32
García-Amaro E (2004) Modificaciones al sistema de clasificación climática de Köppen (para adaptarlo a las condiciones de la República Mexicana). Universidad Nacional Autónoma de México, Mexico City, Instituto de Geografía
García-Arizaga MT, Lugo-Hubp J (2003) El relieve mexicano en mapas topográficos. Instituto de Geografía UNAM, Mexico City, p 16
Geyne RA, Fries C, Probert A (1963) Geología y yacimientos minerales del distrito de Pachuca-Real del Monte. Consejo de Recursos Naturales no Renovables, Mexico City
INEGI (2005) Guía para la Interpretación de Cartografía - Climatológica. Instituto Nacional de Geografía, Estadística e Informática, Aguascalientes, Mexico
INEGI (2010) Compendio de información geográfica municipal 2010. Pachuca de Soto, Hidalgo. Instituto Nacional de Geografía y Estadística, Aguascalientes, Mexico
INEGI (2017) Anuario Estadístico y Geográfico de Hidalgo. Instituto Nacional de Geografía y Estadística, Aguascalientes, Mexico
INEGI (n.d.a) Geografía y medio ambiente: Climatología, Mapa. Instituto Nacional de Geografía y Estadística. https://www.inegi.org.mx/temas/climatologia/#Mapa. Accessed 8 Sep 2022
INEGI (n.d.b) Geografía y medio ambiente: Edafología. Instituto Nacional de Geografía y Estadística. https://www.inegi.org.mx/temas/edafologia/. Accessed 8 Sep 2022
INEGI (n.d.c) Geografía y medio ambiente: Uso de suelo y vegetación. Instituto Nacional de Geografía y Estadística. https://www.inegi.org.mx/temas/usosuelo/. Accessed 8 Sep 2022
Krasilnikov P, Jiménez-Nava FJ, Reyna-Trujillo T, García-Calderón NE (2011) Geografía de los suelos de México. Universidad Nacional Autónoma de México, Mexico City, pp 57–118
Krasilnikov P, Gutiérrez-Castorena MC, Ahrens RJ, Cruz-Gaistardo CO, Sergey S, Solleiro-Rebolledo E (2013) The soils of Mexico. Springer, Heidelberg, pp 61–148
López-Levi L, Valverde-Valverde MC (2015) Pueblos mágicos: breves apuntes desde una visión interdisciplinaria (volumen II). Universidad Autónoma Metropolitana, Mexico City, pp 51–53
Lorenzo-Monterrubio A (1995) Arquitectura, Urbanismo y Sociedad en Pachuca, periodo del Porfiriato. Gobierno del Estado de Hidalgo, Pachuca de Soto, Mexico
Morrison A (2003) Los tranvías de Pachuca. https://dacomedores.tripod.com/pachuca/ Accessed 8 Sep 2022
Ortega-Morel J (2015) Minería y ferrocarriles en la región de Pachuca y Real del Monte durante el Porfiriato. Universidad Autónoma del Estado de Hidalgo, Pachuca de Soto, Mexico
Pavón-Hernández NP, Meza-Sánchez M (2009) Cambio climático en el estado de Hidalgo: clasificación y tendencias climáticas. Universidad Autónoma del Estado de Hidalgo, Pachuca de Soto, Mexico, pp 41–118
Pérez-Torres DE (2013) Las zonas metropolitanas de México: estructuración urbana, gobierno y gobernanza. Universidad Autónoma Metropolitana, Mexico City, p 111
Pontón G (1972) Gran diccionario enciclopédico ilustrado. Grijalbo Mondadori, Barcelona

Probert A (2011) En pos de la plata. Gobierno del Estado de Hidalgo, Pachuca de Soto, Mexico

Riegl A (2017) El culto moderno a los monumentos - Caracteres y origen. Machado Grupo de Distribución, Madrid

SCEH (2017) Título del documento subordinado. In: Plano de 1926 1926-CGE-7246-A, Mapoteca Manuel Orozco y Berra. Secretaría de Cultura del Estado de Hidalgo, Pachuca de Soto, Mexico

SCT (2021) Antecedentes de la Reconstrucción y Conservación de Carreteras. Secretaría de Comunicaciones y Transportes. https://www.sct.gob.mx/carreteras/direccion-general-de-conservacion-de-carreteras/antecedentes/ Accessed 8 Sep 2022

SGM (1997) Carta Geológico-Minera, Pachuca F14–11, Hidalgo, Querétaro, Estado de México, Veracruz y Puebla, escala 1: 250,000. Servicio Geológico Mexicano, Pachuca de Soto, Mexico

SICT (2003) Métodos de muestreo y prueba de materiales M-MMP-1–01/03. Secretaría de Infraestructura, Comunicaciones y Transporte, Mexico City

SOP (1965) Proyectos tipo para obras de drenaje para carreteras. Secretaría de Obras Públicas, Dirección General de Proyectos y Laboratorios, Departamento de Vías Terrestres, Mexico City

Torres-Carral G (2012) Desarrollo compatible: nueva ruralidad y nueva urbanidad. Universidad Autónoma de Chapingo—Plaza y Valdez, Mexico City, pp 175–190

UN (2022) The sustainable development goals report 2022. United Nations. https://desapublications.un.org/publications/sustainable-development-goals-report-2022. Accessed 1 May 2023

UN-Habitat (2021) Urban-rural land linkages: a concept and framework for action. United Nations Human Settlements Programme, Nairobi

Wikiloc (2010) Sendero donde corría el tren Pachuca-Real del Monte. https://es.wikiloc.com/rutas-senderismo/sendero-donde-corria-el-tren-pachuca-real-del-monte-1299345. Accessed 8 Sep 2022

# Chapter 5
# Blue and Green Infrastructure: History and Experiences in Mexico and the Arid and Semi-Arid Global South

**Gabriela A. Vázquez-Rodríguez, Marcelino Antonio Zúñiga-Estrada, and Joyce E. Ortiz-Hernández**

**Abstract** Current urban development poses numerous challenges for the Global South, mainly for its arid and semi-arid regions. First, this chapter aims to present the emerging blue and green infrastructure (BGI) concept as the result of a historical process whose deep roots can be found in the Neolithic and Industrial revolutions, the ecological conservation theory, but also in the traditional knowledge of communities forming diverse socioecological systems in the Global South. Then, as a part of its conceptual framework, which relies heavily on the provision of ecosystem services, the similarities and differences between BGI and other concurrent concepts are presented. Second, we review the ecosystem services provided by BGI, as well as some examples of its implementation in the Global South, emphasizing its benefits in terms of water quantity and quality, and the contribution of BGI to the United Nations Sustainable Development Goals. Finally, although the BGI approach entered only recently into the national policy discourse, we present some governmental, academic, and social endeavors with positive effects on sustainable water management in Mexico. Our view is that the perspective of water conservation implicit in BGI is of urgent implementation in the arid and semi-arid growing cities of the Global South.

G. A. Vázquez-Rodríguez (✉)
Chemistry Academic Area, Autonomous University of Hidalgo State, Carr. Pachuca-Tulancingo Km 4.5, Col. Carboneras. Mineral de La Reforma, C.P. 42184 Hidalgo, Mexico
e-mail: gvazquez@uaeh.edu.mx

M. A. Zúñiga-Estrada
Earth Sciences and Materials Academic Area, Autonomous University of Hidalgo State, Carr. Pachuca-Tulancingo Km 4.5, Col. Carboneras, C.P. 42184 Mineral de La Reforma, Hidalgo, Mexico
e-mail: marcelino_zuniga@uaeh.edu.mx

J. E. Ortiz-Hernández
Iztépetl Ciencias de la Tierra y Afines S.C., Avenida 2 306, Parque de Poblamiento 1a. Secc, Plutarco Elías Calles, C.P. 42032 Pachuca, Hidalgo, Mexico
e-mail: joyce_ortiz@geosphera.mx

 
L. Lizárraga-Mendiola et al. (eds.), *Sustainable Spaces in Arid and Semiarid Zones of Mexico*, The Latin American Studies Book Series,
https://doi.org/10.1007/978-3-031-53418-8_5

**Keywords** Water conservation · Historical development · Industrialization · Urbanization · Ecosystem services · Natural corridors

## 5.1 Introduction

Blue and green infrastructure (BGI) is an integrated and cost-effective approach to urban and landscape planning that is receiving attention as an effective and flexible strategy for mitigating and adapting to disturbances generated by climate change and urbanization, thus promoting urban sustainability and resilience (UNEP 2014). BGI has been defined as an interconnected network of natural and designed landscape components that include the interaction of open green spaces and water bodies. A short list of BGI solutions includes riparian buffers, green roofs, bioretention systems, and constructed wetlands; also, there are comprised devices whose operation, unlike the previously mentioned, does not require the use of vegetation, such as rainwater or fog harvesting systems, sand dams, permeable pavements, or infiltration trenches. In addition to enhancing water security, it has been argued that these components offer services such as flood control, improvement of water quality, and urban aesthetics (UNEP 2014).

Despite the benefits that BGI could bring to the arid and semi-arid zones, studies reporting its full-scale application and effectiveness in these geographies are still lacking (Zúñiga-Estrada et al. 2022). Besides, a recent review points out the scant presence of the Global South countries in the otherwise abundant BGI bibliography (Ying et al. 2022), which should be amended. These countries are currently immersed in a fast urbanization process, so the search for sustainable strategies to limit the negative consequences of this process is urgent.

In this chapter, we briefly revise the global development of infrastructure as the context in which the BGI concept arose, explore the contribution of indigenous and traditional knowledge to water management in the Global South, and present some of the problems resulting from industrialization and urbanization. Finally, we present BGI as the integration of concepts derived from neighboring fields such as urbanism, landscape planning, and conservation ecology, as well as its contribution to enhancing water security and in general to achieving the United Nations Sustainable Development Goals (SDGs), particularly in arid and semi-arid zones worldwide and in Mexico.

## 5.2 Infrastructure Through the Ages: An Overlook

The word "infrastructure" was first used in English in 1927; at that time, it was a French term designating the substrate material below railroad tracks (Carse 2012). Now, it is mainly used to design the socio-technical systems developed from the Neolithic period that allow for the interconnection and movement of people, things, and ideas (Namba 2016). As a part of the built environment, infrastructure designates landscape-scale systems based on elements built, maintained, and used by more than one household's inhabitants (Smith 2016). Commonly, waterworks, electricity grids, roads, and railways, among many other societal endeavors, are referred to as infrastructure. These systems are composed of hard and soft elements: material assets on the one side and the institutions that design, manage, govern, finance, and regulate them on the other side (Chester and Allenby 2018). Infrastructure delivers resources such as energy, water, and information or provides services such as people mobility or waste collection and processing. From these facts, infrastructure has been likened to the veins, arteries, and nodes of living bodies (Christensen 2017).

As already said, the technological, economic, cultural, and social changes that brought about the agricultural development, known as the Neolithic revolution, also marked the infrastructure's onset. Quite likely, wells and roads were the first infrastructure projects. Concerning roads, some nomadic gatherer-hunter groups traveled through trails such as the Old North Trail, already used 10000 years ago by indigenous groups going from the Yukon Territory to New Mexico, or the routes composing the Wixárika path from Nayarit to Wirikuta, in Mexico, that connect natural sacred sites along more than 800 km. However, agricultural communities were those who consistently built roads. For example, Egyptians constructed the Lake Moeris Quarry Road, the oldest existing paved road, in Giza about 2500 years BCE (Lay et al. 2021). Besides, these Neolithic human groups began to deal with challenging landscape features such as ravines or ponds rather than only avoiding them. In England, timber tracks were built over bogs around 4000 BCE (Coles 1989). The Arkadiko Bridge is believed to be the oldest stone bridge still in use. The Myceneans constructed it around 1300–1190 BCE in the Peloponnese, Greece (Zhang 2014).

Infrastructure focused firmly on irrigation in this first historical stage. Sumerian civilization's rise in Mesopotamia, around 4000 BCE, was due to its mastery of hydraulics, which allowed harnessing the water from the Tigris and Euphrates rivers and managing their seasonal floodings. Through the construction of canals, reservoirs, or dykes, Sumerians assured proper water supply to their crop fields and towns, some of which were furnished with wastewater and surface-based stormwater drainage systems (Mays 2010). Ancient Egypt's transition from natural to irrigated agriculture founded on infrastructure occurred toward 3100 BCE (Mays 2010).

The first planned and organized sewerage and drainage systems are attributed to the Minoan and Harappan civilizations, which flourished in Crete and the Indus Valley, respectively (by 3000 BCE). Minoans are thought to have built the first known "flushed" toilets in the Knossos Palace, around 1700 BCE. They were connected to a closed sewer that is still operational (Lofrano and Brown 2010). Harappans developed centralized and decentralized wastewater management in some main sites, such as Harappa and Mohenjo-Daro (De Feo et al. 2014). Wastewater from individual households passed through terracotta pipes into a sump, where a preliminary settling of solids occurred before being conducted to agricultural fields by a network of covered sewers. These systems are among the first known attempts at wastewater treatment (Lofrano and Brown 2010).

These increasingly hierarchized societies transported goods through infrastructure, which began to take place even by sea. Some of the oldest ports known were constructed by the Egypt 4th dynasty (2500 BCE) in Wadi el-Jarf to transport minerals through the Red Sea and by the Harappan civilization (2300 BCE) in Lothal, in the Arabian Sea, to commerce with Mesopotamia (Dwarakish and Salim 2015). Infrastructure was also the material manifestation of a leader's power, as the Processional Way and the Ishtar Gate of ancient Babylon (600 BCE), in Mesopotamia, or the way to extend the influence of a particular authority. Some examples of the latter include the Persian Royal Road, restored under Darius the Great's rule (522–486 BCE), or afterward the Roman roads, a 400000 km network that partially resurfaced the Persian Royal Road and stretched from Northern England to Egypt and beyond at its zenith (toward 100 CE) (Temjanovski et al. 2017). In Mesoamerica, maybe around 250–900 CE, the Mayans built roadways called *sacbeob* (plural of *sacbe*, "white roads"), which are raised stone platforms paved with limestone. The vast and smooth surfaces of sacbeob allowed the movement of people, goods, and water between settlements and water reservoirs, in a region where burden beasts were unknown. Some sacbeob also served to divert or impound water to select locations (Shaw 2012). In South America, the *Qhapaq Ñan* was a 23000 km road system stretching across the Inca Empire, built on one of the roughest terrains of the world (Tomateo 2021). Romans and their subjects transformed the landscape extensively through draining, terracing, or centuriation, among other practices (Franconi 2021). At their height, the famous Rome aqueducts extended over 500 km, of which only 73 km were constructed over the land surface (Mays 2010). Daily, eleven aqueducts, built from the end of the fourth century BCE and over the following 600 years, came to supply around a million inhabitants with one million cubic meters of spring water (Mithen 2012).

In 1435 CE, from an island in the Valley of Mexico basin, the Mexica started the expansion of their empire. In this formerly endorheic basin, which consisted of five interconnected lakes whose levels changed seasonally, a complex hydraulic structure (dikes, aqueducts, canals, and diversion dams, among others) was constructed to create urban settlements and agricultural land, manage floods, maintain freshwater supply, and navigate (Cassin and Ochoa-Tocachi 2021). This hydraulic infrastructure was the basis for sustaining around one million inhabitants in the basin at the peak of the Mexica empire. European ports, such as those of the Hanseatic League, Seville, Cadiz, and Northern Italy, allowed the European expansion toward America, Africa, and Asia, and the following colonialism wave from the fifteenth century onward (Marnot 2020). These ports constituted centers of accumulation of capital and wealth, to the point that 1492 can be considered the foundational moment of modernity and capitalism.[1] Indeed, the decolonial posture considers that European industrialism

---

[1] In the same vein, the Orbis hypothesis proposed by Lewis and Maslin (2015) holds that the dip in atmospheric $CO_2$ observed between 1570 and 1620 in ice core records is the result of the exhaustive carbon sequestration derived from land decultivation that, in turn, followed the arrival of Europeans to America in 1492. This atmospheric $CO_2$ reduction is coincidental with the decimation of three-quarters of American and African populations (Vázquez 2018) and has been proposed as the beginning of the Anthropocene epoch (Lewis and Maslin 2015).

was fueled by the plantation economy, based on land usurpation, enslaved labor, and the traffic of valuable goods obtained at low or no cost from the new colonies (Vázquez 2018).

Globally, agriculture-related infrastructure resulted in substantial shifts in water and land use patterns, resource consumption, local and regional biodiversity, and in the flows of materials such as nutrients, waste, and metals, among others (Chester et al. 2019). Over time, the infrastructure deployment associated with the Neolithic Revolution supported high population growth rates worldwide, leading from five or six million people at the end of the Paleolithic period to 150 million around 1000 BCE. Such a leap in the world population would only be surpassed with the Industrial Revolution, as it passed from 500 million in 1650 to one billion in 1850 (Encyclopedia Britannica n.d.).

### *5.2.1 Water Infrastructure in the Global South*

As was presented above, many societal groups have developed systems to deal with environmental challenges over time. Those still practiced locally or by small communities are commonly referred to as indigenous and local knowledge (ILK). This term encompasses knowledge-practice-value systems, passed from generation to generation, that have emerged from a local understanding of the human–nature relationship and have evolved through adaptive learning (Cassin and Ochoa-Tocachi 2021). Water is often, if not always, a component of ILK systems. In the approach detailed by Lane (2014), hydraulic infrastructures can be classified as wet or dry if water is a constant or temporary component, respectively. One example of wet hydraulic infrastructure is the Mexica *chinampas*[2] ("floating garden"), consisting of artificial islands built in strips with lake sediments, branches, and decaying vegetation, and surrounded by three or four channels connected to ditches and acting as water body and drainage (Government of Mexico City 2017).

Among dry infrastructures, one remarkable example is given by the *qanats*, underground aqueducts constructed in Iran and Iraq to collect groundwater accumulating in the upland and convey it to villages and agricultural fields in the lowlands (Cassin and Ochoa-Tocachi 2021). Table 5.1 presents some historical and current examples of wet and dry ILK-based hydraulic infrastructure implemented in the Global South.

Thanks to wet and dry infrastructures, water storage, and supply were managed, thereby contributing to maintaining or enhancing water security over time. On the one hand, this infrastructure can be purely hydraulic, i.e., consisting only of built structures such as dams or cisterns for harvesting or storing rainfall or runoff. On the other hand, infrastructure relies often on natural processes (such as infiltration and storage in soils and aquifers) for assuring water supply. In this type of infrastructure

[2] *Chinampas* are still practiced in Xochimilco, Mexico City. It is sustainable and one of the most productive agricultural systems, allowing five crops per year.

**Table 5.1** Historical and current examples of wet and dry hydraulic infrastructure implemented in the Global South

| Type | Infrastructure (antiquity) | Description | Country or historical region |
|---|---|---|---|
| Dry | *Qanats* (3000 BCE)* | Network of underground tunnels conveying water from the water table to the lowland surface[a] | Iran*, Iraq, Al-Andalous, Mexico, China |
| | *Chultun* | Bottle-shaped cisterns dug in the ground to harvest rainwater and surface runoff from buildings and ceremonial places[a] | Yucatan Peninsula (Mexico) |
| | *Andenes* (2480–2320 BCE)* | Artificial cultivation terraces constructed on hills or mountains for growing food and mitigating landslides[b] | Central Peru* |
| | *Puna* (prior to the seventeenth century CE) | Coastal stone wells harvesting rainwater and coastal groundwater discharges[c] | Easter Island |
| | *Q'ochas* | Shallow sunken fields irrigated by canals conveying runoff[b] | Highlands of Peru* |
| Wet | *Amunas* | Artificial aquifer recharge system consisting of infiltration canals sowing water, which is harvested downstream[c] | Peru |
| | *Bofedales* | Small silt dams that raise the water level in a wetland, allowing aquifer recharge and the discharge of water downstream in the dry season[c] | Andean highlands |
| | *Chinampas, camellones* (*waru waru*) | Raised crop fields built in areas of waterlogged soils or prone to annual flooding[b] | Mexico, Titicaca Basin, Venezuela |
| | Sand (underground) dams | Structures built across intermittent streamflows allowing the trapping and buildup of sediments behind the dam. Water is extracted by wells in the dry period[c] | Sardinia, Kenya, India, Ethiopia, Mexico, Brazil |

*Possible antiquity or place of origin; [a]Mays (2010); [b]Lane (2014); [c]Cassin and Ochoa-Tocachi (2021)

built elements may exist, but natural elements such as vegetation, soils, sediments, or aquifers have a fundamental role (Cassin and Ochoa-Tocachi 2021).

### *5.2.2 Urban Infrastructure from the Onset of Industrialization*

From the Industrial Revolution, manufacturing, energy production, and large-scale transportation, among others, required the construction of increasingly complex infrastructure. James Watt's steam engine was undoubtedly the icon of the Industrial Revolution, and it was soon adapted to create new transport modes, such as steamships or railway locomotives. Although rail-traveling vehicles were known even in Ancient Greece and Pompeii, they relied on human or animal power, conveyed traffic consigned by their owners, and carried mostly minerals. After Richard Trevithick built the first steam-powered locomotive in 1801, the first railway intended for public passenger transport was inaugurated in England in 1825 (Casson 2009). Railways were, in turn, major drivers of cultural transformations in the nineteenth century: not only did they shorten the time needed to transform industrially raw materials into goods, but also expanded the space where individuals thrive far away from their birthplace.

Electricity was an inquiry subject from antiquity. Its scientific study probably started in 1745, with the invention of the first device able to store large amounts of electric charge, the Leyden jar (Reliance Foundry 2022). After Benjamin Franklin demonstrated, in 1752, that lightning and electricity were the same phenomenon, numerous inventions appeared all through the nineteenth century: the electrical battery, by Alessandro Volta (1800); the electrical motor, by Michael Faraday (1821); the lead-acid battery, by Gaston Planté (1859); the water turbine, by Lester Allan Pelton (1880). All of these, among others, contributed to the starting of the first coal-fired power station generating electricity for public use, in London (1882). The first transmission of AC electric power from Niagara Falls to Buffalo, New York, in 1896, marks the beginning of both world electrification and the Second Industrial Revolution. The innovations in the energy field catapulted in turn the emergence of telecommunications. The electrical telegraph was invented in 1816, but its coupling to a lead-acid battery in 1859 contributed largely to the expansion of this technology. Later, Guglielmo Marconi enabled the wireless transmission of messages through Hertzian waves, boosting the long-distance transmission and reception of messages.

The consequent building and support of more and larger cities led to skyrocketing infrastructure development (Chester and Allenby 2019). Throughout the nineteenth century, the growing industrial workforce required in cities such as New York, Glasgow, or Berlin fueled the emergence of tenements, either by the construction of large apartment buildings, or the division of single-family dwellings into several spaces. If urban infrastructure was previously dictated by environmental constraints (for instance, by scarce freshwater sources in the vicinity), industrialization, technological innovations, and capital accumulation (derived partially from the second wave of colonialism) allowed cities to transcend these constraints in a quest for new resources, at least momentarily.

This first stage of urban development, marked by industrialization and ending around 1910, is also characterized by an expansion of municipal oversight of infrastructure development through committees or departments, and by the establishment of core infrastructure (Porse 2014). Some examples of this core infrastructure are the central water supply (accompanied by early drinking water treatment), and the sewer networks established in several European and USA cities in the late nineteenth century. These endeavors are today known as gray infrastructure and were modernization projects driven by the application of expanding scientific knowledge to urban design. Another distinctive element of this period was the new transportation technologies (steam and electric-powered streetcars), which prompted municipal authorities to build paved roads and bridges (Porse 2014). From the innovations made around 1860 in its production process, steel could be produced massively and became the backbone of most infrastructure projects for which iron hardness was not enough.

The next stage (1910–1950) in urban infrastructure came mainly from the rise of automobile use for personal transit (Porse 2014). In the USA, the growing affordability of automobiles started by the Ford Company encouraged suburban expansion and the construction of multi-lane roads, highways, and secondary arteries enabling people to transport between their homes and industrial or business centers.

Since around 1950, the environment and human societies have gone through a qualitative leap, represented above all by the beginning of the nuclear era, and quantitative hikes in global socioeconomic indicators and Earth trends such as population, water use, fertilizer consumption, atmospheric $CO_2$, or ocean acidification, among many others (McNeill and Engelke 2016). This deflection in social metabolism and anthropogenic environmental impacts became known as the Great Acceleration, and of course, also refers to the sharp upward shift of the urban population. New York and Tokyo became the first megacities, that is, urban agglomerations with over 10 million inhabitants, in the 1950s. Cities expanded radially through the development of outer ring suburbs, while inner neighborhoods shrank and the infrastructure of downtowns tended to be neglected. In New York City, to overcome the degradation and rise of the criminality of inner neighborhoods, massive highways connecting them with peripheral areas were constructed, at the expense of parts of these neighborhoods (Porse 2014). Despite non-stop road and highway construction, usually planned by central and expert-driven committees, traffic congestion appeared in major cities. Other problems related to urbanization developed too. The imperviousness of urban constructions alters the local hydrology because it prevents rainwater infiltration into the soil's lower layers. Simultaneously, runoff increases, thereby diminishing the potential for aquifer recharge. A greater volume of runoff, in turn, leads to more severe and frequent floods, which became increasingly common in urbanizations, even in arid and semi-arid climates (Zúñiga-Estrada et al. 2022).

### 5.2.3 Urban Infrastructure and Environmentalism

In the 1960s, environmentalism emerged as a global countercultural phenomenon. The spread of the idea that humanity should coexist harmoniously with nature rather than seek to dominate it depended on several factors. One of these was that during the preceding decades, pollution had accumulated dangerously in the industrialized world and had already shown its effects. Through her influential book *Silent Spring* (first published in 1962), Rachel Carson revealed to the general public the disastrous effects of the use of organochlorine pesticides and questioned the ethics underlying the extensive use of chemicals that had just been synthesized and whose effects were yet unknown. So, from the 1960s, environmental protection was incorporated into the concerns of taxpayers and thus into government agendas. Environmentalism gained more momentum with the publication of *Limits of Growth* (1972), the report of the Club of Rome that examined possible scenarios of population growth, food production, industrialization, use of non-renewable natural resources, pollution, and the planet's carrying capacity. Although several of its predictions proved incorrect, the report contributed to changing how the society–environment relationship was perceived.

The publication of *Our common future*, also known as the Brundtland report, in 1987, first defined sustainable development as "the development that meets the needs of the present without compromising the ability of future generations to meet their own needs". Since then, governments, enterprises, and social institutions have adopted sustainable development as their new paradigm, leading to policies and regulations intending to prevent further environmental degradation. All the while, there was growing scrutiny of the technocratic elitism that overlooked local contexts and cultures and that sometimes had the form of classist or racist infrastructural projects. A classic example is the bridges on the highways connecting New York City to Long Island beaches, which were planned by Robert Moses (the "Master Builder" of New York City) and constructed low enough to avoid buses from passing underneath. This design decision excluded common users of public transport, that is, racial minorities and low-income groups, and favored upper and middle-class white people who owned cars (Winner 1986). This type of centrally planned infrastructure that also divided communities began to face opposition from local stakeholders pursuing self-organization and neighborhood revitalization (Porse 2014).

## 5.3 The Rise of Blue and Green Infrastructure

BGI is an old idea behind a new term, that is the result of the integration of concepts derived from neighboring fields such as urbanism, landscape planning, and conservation ecology. In the nineteenth century, on the one hand, urbanism was emerging as a rational discipline necessary to the modern cities and, on the other hand, the Romantic movement was calling for a return to the ideals of beauty and liberty embodied in the first place by nature, and in the second place by the models of ideal cities conceived in the Renaissance period. Some seminal propositions can be traced in Haussmann's unrealized plan for connecting the peripherical forests of Paris into a green belt around the urban tissue (Hrdalo et al. 2021), the Boston Emerald Necklace and greenways proposed in 1880 and 1887, respectively, by Frederick Law Olmsted in the USA, and the Garden City movement inspired by the British Ebenezer Howard. In contrast to the Haussmann plan, greenways and garden cities were actually developed in several European and USA cities as complementary to each other (Mirici 2022). These concepts also materialized in the Green Belt and the Finger Plan projected for London (1935) and Copenhagen (1947), respectively, seeking to rationally organize the urban sprawl. Other European cities such as Zagreb, Helsinki, Amsterdam, and Freiburg had similar projects (Mirici 2022).

In particular, the green belt concept enjoyed certain popularity worldwide, as it was transmitted from London to other countries through the British colonial channels (Amati and Taylor 2010). For instance, in Jerusalem, British authorities instigated the creation of a green belt before and after the Second World War to control the Palestinian spread (Amati and Taylor 2010). Later, during the second half of the twentieth century, corridors linking parks, meadows, forested areas, urban ponds, and waterways, among other elements, were developed in some USA cities, notably in Philadelphia, Boston, and Washington (Hrdalo et al. 2021). The enhancement of ecological connectivity derived from enabling these corridors is a valuable element incorporated later in the BGI concept.

Another BGI theoretical foundation is the idea of ecosystem services. Although this concept has a deep history, an immediately preceding idea is the "ecosystem function" formulated by Eugene Odum in 1956 to design the processes operating within an ecological system irrespectively of their usefulness for humans. As presented in the preceding section, environmental concerns were consistently incorporated into public policies, institutions, education, and the mainstream discourse from around 1960. Environmental impacts of infrastructure, industrial activities, and housing, among others, became as tangible as the benefits they have brought. At the same time, the benefits that natural ecosystems offer began to be recognized. In the 1960–1970 decade, the notion of the usefulness of nature for society was addressed by natural scientists, who used terms such as "functions of nature" to denote the ecosystem provision of work, space, and goods to human societies, or "natural capital" to design natural resources that should be treated as capital items of value rather than income (Braat and de Groot 2012). The ecosystem services theory[3] was fully applied for evaluating economically these services on a global scale in 1997 and formalized in 2005 with the publication of the United Nations Millennium Ecosystem Assessment (Mirici 2022). Ecosystem services were grouped into four main categories: provisioning services (production of food, fuel, and fibers); regulation services (climate maintenance, regulation of floods and diseases, pest biological control, crop pollination, and water purification); cultural services (provided as pharmaceutical products or educational, recreational, and spiritual benefits); and finally supporting services enabling all the above (Boehnert 2016).

Green infrastructure (GI) was first defined by Benedict and McMahon (2002) as "an interconnected network of green space that conserves natural ecosystem values and functions and provides associated benefits to human populations". The emergence of the GI concept prompted academics and practitioners of several European and USA cities to discuss the GI worth in their specific contexts, which laid the foundations for site applications. Subsequent works helped to disseminate the potential human health benefits of GI and its role in adapting cities to climate change, thereby promoting investment in GI projects and engaging policy debates (Mell

[3] The concept of ecosystem services is however problematic. It has been said that is a reductionist approach that instrumentalizes ecological relations, where ecosystem services are wrongly considered as a component of market and not the possibility for markets to exist. One of the major criticisms is that its application in the EU Emissions Trading Scheme has not reduced gross carbon emissions (Boehnert 2016).

2017). Although in some definitions its application to rural settings is envisaged, GI is mostly used in the context of urban landscape planning at multiple scales (e.g., landscape, region, or site scale) and across jurisdictional limits.

Some GI strategies include reforestation, forest conservation, reconnection of rivers with floodplains, and conservation or restoration of grassland, prairies, and natural wetlands. Other GI solutions are designed systems such as constructed wetlands, bioswales, rain gardens, or green roofs, whose effectiveness in providing an ecosystem service (as the mitigation of urban runoff) is largely based on the presence of vegetation (Fig. 5.1). Unvegetated (or blue) solutions, such as rainwater and fog harvesting systems, infiltration wells, or permeable pavements, rely on natural or built elements for water conservation and management but are entirely compatible with GI philosophy. For this reason, blue-green or blue and green infrastructure (BGI) is an umbrella term that is gaining popularity along with GI. Connectivity of BGI solutions within a particular scale of application is key to achieving the potential benefits of this approach.

**Fig. 5.1** Vegetated area in the National Autonomous University of Mexico (Mexico City) adapted as a rain garden for urban runoff control

### 5.3.1 Underlying Philosophy of Blue and Green Infrastructure and Concurrent Concepts

BGI concept relies on the recognition that natural ecosystems are just as valuable, if not more, as gray infrastructure for societies, and that the non-built space is an indispensable urban structuring element. Conversely, it places urban landscape planning as a conservation strategy, which before the emergence of the BGI concept had focused on the preservation of wildlife and the protection of certain areas. In this way, BGI is a powerful approach that searches for overcoming deeply held dichotomies such as rural/urban, built/non-built, natural/artificial, among others.

Natural infrastructure and nature-based solutions are two concurrent concepts that were coined more recently than BGI. Natural infrastructure is the network of natural spaces with different degrees of human alteration that preserve the values and functions of ecosystems and therefore provide several services to societies. It is understood as a non-antagonistic complement to gray infrastructure that includes BGI (Ochoa-Tocachi et al. 2022). Nature-based solutions are a concept promoted and defined by the International Union for the Conservation of Nature as "actions to protect, sustainably manage and restore natural or modified ecosystems that address societal challenges effectively and adaptively, simultaneously providing human well-being and biodiversity benefits" (Cohen-Shacham et al. 2016). These two concepts are broader than BGI as they are less centered in the urban context; nevertheless, all three are often used exchangeably.

### 5.3.2 Blue and Green Infrastructure and Water

Water security is maybe the most crucial challenge faced by current cities, which are also threatened by population growth, urbanization rate, and climate change. To address it, the US Environmental Protection Agency and the United Nations Environmental Program, among other national and international organizations, have been promoting BGI as a holistic approach to water resources management (Mell 2017).

BGI implementation protects natural or managed ecosystems that provide water-related services, and this enhances, increases, or in some cases replaces the ecosystem services provided by gray infrastructure alone (Echavarría et al. 2015). Ecosystem services related to water include water provisioning (for human and non-human use, and aquatic habitat provision); water quality regulation (water purification, erosion prevention, disease control, and water temperature control); moderation of extreme events (riverine flood protection, urban runoff, and coastal flood control), and recreation and other cultural services (UNEP 2014).

Coupling of gray and blue–green infrastructure is increasingly seen as the most cost-effective way of achieving sustainable management of urban water, as it is based on a water conservation perspective able to provide several of the services mentioned above, such as sustaining water supply, reducing runoff pollution, and enhancing climate resilience by moderating extreme climate events, to name but a few. Table 5.2 shows BGI solutions applicable to urban sites (but also at larger scales), along with the primary water-related services they provide, and the conventional gray infrastructure providing the same services (UNEP 2014).

More specifically, the incorporation of BGI solutions to urban environments restores the pre-development hydrology through the enhancement of evapotranspiration, infiltration, and aquifer recharge. Likewise, BGI strategies and solutions improve urban runoff quality through adsorption, filtration, biodegradation, and phytoremediation, which are depollution mechanisms occurring naturally when prolonged contact between water, vegetation, rock aggregates, and soil is allowed (Zúñiga-Estrada et al. 2020).

### *5.3.3 Blue and Green Infrastructure and Sustainable Development*

From the provision of ecosystem services such as the management of urban runoff at its source, the mitigation of nutrient loading, and the reduction of flood impacts, the BGI contribution to transform human settlements in safe, sustainable and resilient places, and consequently to the SDG 11 –Sustainable Cities and Communities– is the most acknowledged (Shackleton 2021). However, as exposed above, BGI provides a wide array of ecosystem services that impact positively on several aspects of human well-being, and so it contributes to other SDG. In the first place, the aforementioned services are closely related to water conservation practices and control of non-point pollution that contribute directly to SDG 6 –Clean Water and Sanitation– and SDG 14 –Life below Water–. Due to the reliance of BGI on urban greenery, its adoption reduces atmospheric carbon dioxide, expand carbon sinks and decrease the urban heat island effect, which is clearly associated to the SDG 13 –Climate Action–, besides of benefiting recreation, physical and mental health, which are under the scope of SDG 3 –Good Health and Well-Being–; through the protection, restoration, or promotion of natural corridors enhancing rural–urban connectivity, BGI also aids in the achievement of targets related to SDG 15 –Life on Land– notably by managing forests, fighting against desertification, reversing land degradation, creating habitat, and enhancing biodiversity (Anderson and Gough 2022). In fact, some authors have argued that the sole effort of tree planting in urban areas helps in achieving 15 of the 17 SDGs (Shackleton 2021).

**Table 5.2** Water-related services provided by blue and green infrastructure (BGI) strategies or solutions applicable to urban contexts

| Primary water-related service to be provided | | BGI strategy/ solution | Location | | | Gray infrastructure solution |
|---|---|---|---|---|---|---|
| | | | Watershed | Floodplain | Urban | |
| Water supply regulation (and drought mitigation) | | Wetlands restoration/ conservation | X | X | X | Dams and groundwater pumping; water distribution systems |
| | | Wetlands construction | X | X | X | |
| | | Water harvesting | X | X | X | |
| | | Green spaces* | | | X | |
| | | Permeable pavements | | | X | |
| Water quality regulation | Water purification | Wetlands restoration/ conservation | X | X | X | Water treatment plant |
| | | Wetlands construction | X | X | X | |
| | | Green spaces* | | | X | |
| | | Permeable pavements | | | X | |
| | Biological control | Wetlands restoration/ conservation | X | X | X | Water treatment plant |
| | | Wetlands construction | X | X | X | |
| | Water temperature control | Wetlands restoration/ conservation | X | X | X | Dams |
| | | Wetlands construction | X | X | X | |
| | | Green spaces** | | | X | |
| Moderation of extreme events (floods) | Riverine flood control | Wetlands restoration/ conservation | X | X | X | Dams and levees |
| | | Wetlands construction | X | X | X | |
| | Urban runoff | Green roofs | | | X | Urban runoff infrastructure |
| | | Green spaces* | | | X | |

(continued)

**Table 5.2** (continued)

| Primary water-related service to be provided | | BGI strategy/ solution | Location | | | Gray infrastructure solution |
|---|---|---|---|---|---|---|
| | | | Watershed | Floodplain | Urban | |
| | | Water harvesting | X | X | X | |
| | | Permeable pavements | | | X | |

Modified from UNEP (2014)
*Bioretention and infiltration; **Shading of waterways

### 5.3.4 *Blue and Green Infrastructure in the Arid and Semi-Arid Global South*

Available BGI applications, assessments, and guidelines have been developed mostly for mesic or temperate climates. However, arid and semi-arid regions present characteristics that must be taken into account in any BGI development. For instance, in these areas, rainfall events tend to be short-lived and random, and then the vegetation cover used in BGI should be endogenous and maybe xerophile to cope with water intermittency (Fig. 5.2). Concerning quality, urban runoff has higher pollutant concentrations in arid and semi-arid zones than for international data, which are typically biased toward mesic zones (Zúñiga-Estrada et al. 2022). So, to be functional in arid and semi-arid zones, BGI has to respond to their particularities and not only be transplanted from other climatic conditions. In the dry regions of the Global South, in addition, consistent hydrological data sets may be infrequent, which results in scarce hydrologically-based BGI case studies. The shortage of good practice examples of BGI, in turn, limits the generation of specific design standards and manuals (Zúñiga-Estrada et al. 2022). Specifically in arid and semi-arid zones, BGI has helped mitigate runoff flows, at least according to hydrological modeling, and does not seem to impair water quality (Meerow et al. 2021). Actually, the removal of some pollutants (suspended solids, metals, and nutrients) from urban runoff has been reported (Zúñiga-Estrada et al. 2022). However, the need for more empirical BGI monitoring and validation has also been pointed out for these climatic regions (Meerow et al. 2021).

International organizations such as the United Nations and the Intergovernmental Panel on Climate Change have acknowledged the crucial role of indigenous peoples in advancing climate and biodiversity solutions. For these people, such solutions are not new but rather an integral part of their survival for centuries. For example, even though indigenous peoples represent around 5% of the world population, they alone are responsible for protecting 80% of the global biodiversity and caring for 20% of the Earth's surface (UN-Water 2018). That is why, to cope with water variability associated with climate change, an alignment of BGI strategies with ILK is proclaimed

**Fig. 5.2** A low area in a semi-arid city (Chihuahua, Mexico) adapted as a rain garden with local and xerophile vegetation

as necessary, along with the addressing of the socio-economic vulnerabilities of the indigenous peoples and the respect for their rights (UN-Water 2018).

In this regard, specifically for Colombia, Ecuador, Peru, and Mexico, the application of their precolonial infrastructure has been recommended to improve water security and community-based sustainable water management, notably concerning the Andean hydraulic terracing system existing in Ecuador and Peru (Echavarría et al. 2015). For these high-mountain ecosystems, one remarkable initiative is the Natural Infrastructure for Water Security Project, which has evaluated the potential contributions of pre-Inca water infiltration infrastructure for improving water security (Ochoa-Tocachi et al. 2019). One important result of the project was that the dry-season flows to Lima city could increase by around 7.5% through the upscaling of an indigenous system that, during the wet season, diverts water from headwater streams onto mountain slopes to enhance downslope natural springs. The project results have also stressed the need for appropriate hydrological modeling to evaluate the benefits and risks of natural infrastructure interventions, which resulted in the publication of detailed guidelines (Ochoa-Tocachi et al. 2022).

Yet, various criticisms have been leveled against the extractive use, or in other words, only serving for scholar production, of this ancestral knowledge (Tomateo 2021). In the Global South, infrastructure has been instrumental to colonialist purposes looking for controlling nature and by extension to control peoples; in Latin

America, nothing could illustrate better this than the Panama Canal (Carse 2012). Tomateo (2021) considers that, due to its intrinsic relationship with nature and politics, BGI can constitute a decolonization tool. To this end, it is essential to democratizing the decision-making process for co-designing and building multifunctional and inclusive BGI. The author also points out that a decolonizing BGI (re)introduces different forms of governance, communal ownership of land, de-standardization, and scalar fluidity, among other disruptive policy practices (Tomateo 2021).

## 5.4 Blue and Green Infrastructure in Mexico

For some years now, the BGI concept has circulated in Mexico as a way to integrate economic development and urban planning with climate change agendas. A unifying vision at the national level, positioning BGI as a risk mitigation and urban water resilience building strategy, is still lacking (SEDEMA n.d.); however, some concrete actions have been carried out. For example, a roadmap to implement BGI was developed as a mitigation and adaptation strategy to climate change in Mexican cities (SEDATU/SEMARNAT/GIZ 2018). Taking this document as a framework, in 2019 the Office of Environment of Mexico City financed the execution of the first stage for the participatory elaboration of the Green Infrastructure Master Plan for Mexico City; the second stage of this plan began in 2020 and, based on the information gathered in the first stage, is generating strategic axes, goals, actions, and indicators, as an initial project portfolio and proposals for funding sources (SEDEMA n.d.). Likewise, the Department of Agrarian, Territorial and Urban Development (SEDATU for its initials in Spanish), in collaboration with the Inter-American Development Bank, prepared the *Manual of streets: road design for Mexican cities* to serve as a reference in the geometric design of streets and the management of road projects in urban areas. In this manual, BGI is contemplated, and guidelines for its construction are offered (SEDATU/BID 2019).

At the local level, there are also initiatives to apply BGI. The first of these consisted of an implementation guide in Los Pedregales's zone of Mexico City to address problems of water supply, drainage management, mobility, and biodiversity conservation, through the recovery of ecosystem services (Suárez et al. 2011). One of the recommendations of the guide is to multiply small-scale interventions (in green walls, corners, or ridges, for example) in the area mentioned before, rather than undertake large-scale works. In Mexico City, the interventions carried out in Los Coyotes Skate Park, the Legaria rain garden, the Granadas Community Center, and the San Juan de Aragón Forest, among others, illustrate the application of BGI. In addition, between 2017 and 2020, the municipal planning institutes (IMPLAN for its initials in Spanish) of three cities (Mérida, Hermosillo, and León) published their BGI management manuals.

In particular, the IMPLAN Hermosillo manual was developed with the support of the Border Environment Cooperation Commission and the North American Development Bank (IMPLAN-Hermosillo 2017). During its preparation, training workshops were held and pilot interventions were carried out in border municipalities to promote its application. Although it was initially conceived as a response to the needs of border municipalities with arid climates, it ended up in an instrument applicable to all municipalities in Mexico. In this way, it is the only manual that describes in detail the best international practices and the implementation techniques of BGI in the Mexican territory, such as microbasin and rain gardens, permeable pavements, or green roofs, among other devices (IMPLAN-Hermosillo 2017). Thanks to this manual, a specific technical standard for BGI was developed for the municipality of Hermosillo (Gobierno del Estado de Sonora 2018). The criteria established, both in the manual and in the standard, have been applied in some of the projects of the city. Besides, this organization published a vegetation guide that can be used in BGI projects in the same municipality (IMPLAN-Hermosillo 2020).

Nonetheless, to take advantage of BGI's growing funding opportunities, it is crucial to determine the expected benefits for the involved communities and ecosystems (Ochoa-Tocachi et al. 2022). For this purpose, it is crucial to accompany BGI design and validation with adequate hydrological modeling and monitoring studies, which are still scarce.

## 5.5 Conclusions

By tracing back infrastructure history in general and BGI history in particular, the breaking points known as the Neolithic and Industrial revolutions were highlighted as important shifts in water and land use patterns and infrastructure development. The decolonial perspective, considering that Industrial Revolution (and the ensuing global urbanization) was fueled by the plantation economy occurring in the American colonies from 1492, was presented, besides some examples of socially inequitable infrastructure developed in the twentieth century before the rise of environmentalism and BGI.

BGI has the potential to benefit the urban arid and semi-arid zones of the world, and to contribute to several of the United Nations SDGs, notably through flood attenuation during extreme weather events and water quality improvement, but also through the mitigation of the climate change impacts and several environmental and health advantages for communities. However, the BGI application in these regions is scarce, as well as the validation of its performance through empirical monitoring and hydrological modeling. These benefits could extend to the peri-urban areas, thereby enhancing urban–rural connectivity, by protecting or restoring the natural corridors implicit in the BGI notion. Everywhere it is installed, BGI has to respond to the particularities of these zones and not only be transplanted from other climatic contexts. The need of incorporating the experiences and indigenous local knowledge grown out of the Global South countries was stressed, along with the importance of decolonizing BGI through the democratization of the decision-making process, the co-design, and building of multifunctional and inclusive community-based practices, including the fostering and rehabilitation of vernacular architecture. It is our view that relevant BGI research and application adapted to local contexts might contribute to the urban resilience of Mexico and the arid and semi-arid Global South.

## References


Amati M, Taylor L (2010) From green belts to green infrastructure. Plan Pract Res 25(2):143–155

Anderson V, Gough WA (2022) Enabling nature-based solutions to build back better—an environmental regulatory impact analysis of green infrastructure in Ontario. Canada. Buildings 12(1):61

Benedict MA, McMahon ET (2002) Green infrastructure: smart conservation for the 21st century. Renew Resour J 20(3):12–17

Boehnert J (2016) The green economy: reconceptualizing the natural commons as natural capital. Environ Comm 10(4):395–417

Braat LC, de Groot R (2012) The ecosystem services agenda: bridging the worlds of natural science and economics, conservation and development, and public and private policy. Ecosyst Serv 1(1):4–15

Carse A (2012) Nature as infrastructure: making and managing the Panama Canal watershed. Soc Stud Sci 42(4):539–563

Cassin J, Ochoa-Tocachi BF (2021) Learning from indigenous and local knowledge: the deep history of nature-based solutions. In: Cassin J, Matthews JH, López-Gunn E (eds) Nature-based solutions and water security—an action agenda for the 21st century. Elsevier, San Diego, pp 283–335

Casson M (2009) The world's first railway system: enterprise, competition, and regulation on the railway network in Victorian Britain. OUP, Oxford, p 1

Chester MV, Allenby B (2018) Toward adaptive infrastructure: flexibility and agility in a non-stationarity age. Sustain Resilient Infrastruct 4(4):173–191

Chester MV, Markolf S, Allenby B (2019) Infrastructure and the environment in the Anthropocene. J Ind Ecol 23:1006–1015

Chester MV, Allenby B (2019) Infrastructure as a wicked complex process. Elementa 7(21)

Christensen PH (2017) Germany and the Ottoman railways: art, empire, and infrastructure. Yale University Press, New Haven

Cohen-Shacham E, Walters G, Janzen C, Maginnis S (eds) (2016) Nature-based solutions to address global societal challenges. IUCN, Gland

Coles JM (1989) The world's oldest road. Sci Am 261(5):100–107

De Feo G, Antoniou G, Fardin HF, El-Gohary F, Zheng XY, Reklaityte I, Butler D, Yannopoulos S, Angelakis AN (2014) The historical development of sewers worldwide. Sustainability 6(6):3936–3974

Dwarakish GS, Salim AM (2015) Review on the role of ports in the development of a nation. Aquat Procedia 4:295–301

Echavarría M, Zavala P, Coronel L, Montalvo T, Aguirre LM (2015) Infraestructura Verde en el Sector de Agua Potable en América Latina y el Caribe: Tendencias. Retos y Oportunidades. Ecodecisión, Forest Trends, The Nature Conservancy

Encyclopedia Britannica (n.d.) Modernization—population change. https://www.britannica.com/topic/modernization/Population-change. Accessed May 20 2022

Franconi TV (2021) The environmental imperialism of the Roman Empire in Northwestern Europe. In: Erdkamp P, Manning JG, Verboven K (eds) Climate change and ancient societies in Europe and the near east. Palgrave Macmillan, Cham, pp 321–345

Gobierno del Estado de Sonora (2018) Norma técnica que establece las características y requerimientos para la infraestructura verde en el municipio de Hermosillo. Boletín Oficial, tomo CCII, número 26, sección II
Government of Mexico City (2017) Chinampa Agricultural System of Mexico City - A Proposal for Designation as Globally Important Agricultural Heritage Systems (GIAHS). https://www.biopasos.com/biblioteca/Chinampa-agricultural-system-mexicocity.pdf. Accessed 6 May 2022
Hrdalo I, Peroković P, Tomić Reljić D (2021) Historical development of urban green infrastructure and possibilities of its implementation in the Republic of Croatia. Prostor 29(1):56–71
IMPLAN-Hermosillo (2017) Manual de Lineamientos de Diseño de Infraestructura Verde para Municipios Mexicanos. Instituto Municipal de Planeación Urbana y del Espacio Público, Hermosillo
IMPLAN-Hermosillo (2020) Paleta Vegetal Hermosillo. Instituto Municipal de Planeación Urbana y del Espacio Público, Hermosillo
Lane K (2014) Water technology in the Andes. Encyclopedia of the History of Science, Technology, and Medicine in non-Western cultures, 2nd ed. Springer, Dordrecht p 1–24
Lay M, Metcalf J, Sharp K (2021) Paving our ways: a history of the world's roads and pavements. CRC Press, Boca Raton, p 7
Lewis S, Maslin M (2015) Defining the Anthropocene. Nature 519:171–180
Lofrano G, Brown J (2010) Wastewater management through the ages: a history of mankind. Sci Total Environ 408(22):5254–5264
Marnot B (2020) Ports as Tools of European Expansion. Encyclopédie d'Histoire Numérique de l'Europe. https://ehne.fr/en/node/12437. Accessed 23 Jun 2022
Mays L (ed) (2010) Ancient water technologies. Springer, Dordrecht
McNeill JR, Engelke P (2016) The great acceleration: an environmental history of the Anthropocene since 1945. Harvard University Press, Cambridge
Meerow S, Natarajan M, Krantz D (2021) Green infrastructure performance in arid and semi-arid urban environments. Urban Water J 18(4):275–285
Mell IC (2017) Green infrastructure: reflections on past, present and future praxis. Landsc Res 42(2):135–145
Mirici ME (2022) The ecosystem services and green infrastructure: a systematic review and the gap of economic valuation. Sustainability 14(1):517
Mithen S (2012) Thirst: water and power in the ancient world. Harvard University Press, Cambridge
Namba M (2016) Becoming a city: infrastructural fetishism and scattered urbanization in Vientiane, Laos. In: Harvey P, Jensen C, Morita A (eds) Infrastructures and social complexity. Routledge, New York, pp 94–104
Ochoa-Tocachi BF, Bardales JD, Antiporta J, Pérez K, Acosta L, Mao F, Zulkafli Z, Gil-Ríos J, Angulo O, Grainger S, Gammie G (2019) Potential contributions of pre-Inca infiltration infrastructure to Andean water security. Nature Sustain 2(7):584–593
Ochoa-Tocachi BF, Cuadros-Adriazola J, Arapa-Guzmán E, Aste-Cannock N, Ochoa-Tocachi E, Bonnesoeur V (2022) Guía de modelación hidrológica para la infraestructura natural. Forest Trends, Lima
Porse EC (2014) Old Solutions and New Problems: On the Evolution of Urban Water Infrastructure and Environments. Ph.D. Dissertation, University of California, Davis
Reliance Foundry (2022) The history of infrastructure. https://www.reliance-foundry.com/blog/history-of-infrastructure. Accessed May 20 2022
SEDATU/BID (2019) Manual de calles: diseño vial para ciudades mexicanas. Secretaría de Desarrollo Agrario, Territorial y Urbano, Banco Interamericano de Desarrollo
SEDATU/SEMARNAT/GIZ (2018) Implementación de infraestructura verde como estrategia para la mitigación y adaptación al cambio climático en ciudades mexicanas. Secretaría de Desarrollo Agrario, Territorial y Urbano, Secretaría de Medio Ambiente y Recursos Naturales y Gesellschaft für Internationale Zusammenarbeit, Ciudad de México

SEDEMA (n.d.) Infraestructura verde. Secretaría del Medio Ambiente, Gobierno de la Ciudad de México. https://www.sedema.cdmx.gob.mx/programas/programa/infraestructura-verde. Accessed 31 Aug 2021

Shackleton CM (2021) Urban green infrastructure for poverty alleviation: evidence synthesis and conceptual considerations. Front Sustain Cities 3:710549

Shaw JM (2012) Roads to ruins: the role of Sacbeob in ancient Maya society. In: Alcock SE, Bodel J, Talbert RJ (eds) Highways, byways, and road systems in the pre-modern world. John Wiley & Sons, Oxford, pp 130–132

Smith ML (2016) Urban infrastructure as materialized consensus. World Archaeol 48(1):164–178

Suárez A, Camarena P, Herrera I, Lot A (2011) Infraestructura verde y corredores ecológicos de los pedregales: ecología urbana del sur de la ciudad de México. Universidad Nacional Autónoma de México, Ciudad de México

Temjanovski R, Dimitrova J, Arsova M (2017) The roads as civilization symbol or brain drain booster: culture changes in spirit of global challenges. J Econ Univ Goce Delcev 2(1):11–23

Tomateo C (2021) Indigenous land systems and emerging of green infrastructure planning in the Peruvian coastal desert: tensions and opportunities. J Environ Policy Plan 23(5):683–700

UNEP (2014) Green infrastructure guide for water management: ecosystem-based management approaches for water-related infrastructure projects. United Nations Environment Programme, International Union for Conservation of Nature, World Resources Institute, The Nature Conservancy

UN-Water, (2018) The United Nations World water development report 2018: nature-based solutions for water. UNESCO, Paris

Vázquez R (2018) El museo, decolonialidad y el fin de la contemporaneidad. Otros Logos. Rev Estud Crít 9:46–61

Winner L (1986) The whale and the reactor: a search for limits in an age of high technology. University of Chicago Press, Chicago

Ying J, Zhang X, Zhang Y, Bilan S (2022) Green infrastructure: systematic literature review. Econ Res Ekon Istraž 35(1):343–366

Zhang ZJ (2014) Research on the developing process and the types of bridges. Appl Mech Mater 501:1117–1120

Zúñiga-Estrada MA, Vázquez-Rodríguez GA, Ramírez-Cardona M, Lizárraga-Mendiola L (2020) Mineralogical evidence of unfavorable geochemical interactions between urban runoff pollutants and filter materials in low-impact systems. J Hazard Mater 390:122136

Zúñiga-Estrada MA, Lizárraga-Mendiola L, Bigurra-Alzati CA, Aldana-Alonso SE, Ramírez-Núñez JS, Vázquez-Rodríguez GA (2022) Preliminary model-based evaluation of water conservation strategies in a semi-arid urban zone. Land 11(1):101

# Chapter 6
# Traditional Construction Systems and Their Sustainable Opportunities: Adobe and Rammed Earth

**Luis Daimir López-León, Alejandro González-de-la-Fuente, and Liliana Lizárraga-Mendiola**

**Abstract** This book chapter presents an eco-construction process that aims to promote sustainable building practices in arid and semi-arid zones. The proposed construction technique utilizes adobe and rammed earth methods, which are both environmentally friendly. To establish the feasibility of the construction technique, the chapter identifies the geotechnical and mineralogical characteristics of the two materials. Subsequently, the manufacturing and construction processes and their thermal, acoustic, and mechanical properties are studied. The analysis shows that the proposed eco-construction practice is sustainable, with excellent durability and a positive environmental impact. Additionally, it was found that using adobe and rammed earth techniques for construction aligns with some of the United Nations' Sustainable Development Goals (11 and 12). These techniques use natural resources that are readily available locally, reducing the construction process's environmental impact. Furthermore, the chapter evaluates reusable and recyclable materials as a preservation strategy for local ecosystems. It is worth noting that climatic characteristics and geographical conditions play a crucial role in the effectiveness of these construction techniques. Proper maintenance is also essential to extend the durability of the structures. By adopting these eco-construction practices, the construction industry can contribute to the sustainable development of communities and ecosystems in arid and semi-arid zones.

**Keywords** Eco-building · Mud blocks · Rammed earth · Vernacular architecture

L. D. López-León (✉) · A. González-de-la-Fuente · L. Lizárraga-Mendiola
Engineering and Architecture Academic Area, Autonomous University of Hidalgo State, Carr. Pachuca-Tulancingo Km 4.5, Col. Carboneras, C.P. 42184 Mineral de La Reforma, Hidalgo, Mexico
e-mail: luis_lopez@uaeh.edu.mx

A. González-de-la-Fuente
e-mail: alejandro_gonzalez@uaeh.edu.mx

L. Lizárraga-Mendiola
e-mail: mendiola@uaeh.edu.mx

 
L. Lizárraga-Mendiola et al. (eds.), *Sustainable Spaces in Arid and Semiarid Zones of Mexico*, The Latin American Studies Book Series,
https://doi.org/10.1007/978-3-031-53418-8_6

## 6.1 Urban Growth and the Need for Sustainable Edification

Environmentally speaking, the construction industry accounts for 30% of carbon dioxide emissions. It consumes more virgin materials (about 3000 Mt/year) than any other economic activity, being an unsustainable industry (Pacheco-Torgal and Jalali 2012). Using local materials reduces the construction's carbon footprint since the need to transport materials over long distances is minimized. In addition, using natural and renewable materials, such as wood, clay, adobe, stone, or bamboo, reduces the energy needed to manufacture and process construction materials. As urban growth continues, the demand for natural resources to build more infrastructure augments, affecting the natural equilibria. Among the main advantages of an economically sustainable metropolitan area is the efficient and fair distribution of economic resources. Hence, there is a pressing need for affordable housing for low- and middle-income residents (Martínez-Bravo et al. 2019). This is especially relevant in nations with unstable economic development and inferior to developed countries, such as the USA and Canada, in America, South Korea, Japan in Asia, or the European continent. In developing countries such as Mexico, urbanization proliferates in informal settlements where the population cannot afford quality housing with adequate services. Therefore, it is essential to ensure that residents of Mexico's arid and semi-arid areas have access to safe, affordable, accessible, and sustainable housing that allows them to live healthy and prosperous lives (Sustainable Development Goal, SDG, target 11.1, UN 2022).

### *6.1.1 Varieties of Eco-Buildings and Their Benefits*

Eco-building processes possess excellent characteristics that assure good performance; they can be manufactured, used, recycled, or disposed of while having a low environmental impact and being kind to humans (Oyawa 2018). The following are some existing types of materials employed. Wood is easily reusable, offers excellent thermal insulation and noise protection; it is antiseismic and fire-resistant. Cob can be elaborated with clay and sand; it is antiseismic, used in colder areas, and receives solar energy benefits. Bamboo is used due to its resistance, great strength, and flexibility; it reduces hydrogeological instability and atmospheric pollution. Straw bales have very high insulating properties, are fire-resistant and easily recyclable, have excellent thermal insulation properties, and require little manufacturing energy. Adobe is mainly used in arid and semi-arid areas where other natural resources such as wood or bamboo are scarce. Its construction process consumes low amounts of water and produces shallow waste; for hot climates, its thermal properties allow it to maintain cool temperatures during the day and warm at night (Christoforou et al. 2016). Rammed earth is an ancient form of construction, usually associated with arid areas; it provides good tolerance to moisture and rain, is a recyclable material with remarkable acoustic characteristics, and is highly malleable.

These environmental benefits can be even more significant if other options are explored, such as incorporating waste from different local sources to improve the conventional properties of adobe and rammed earth. These benefits will be explored in a later section.

### 6.1.2 Construction with Earth Materials

Despite the variety of the raw resources mentioned above, not all are suitable for adapting to every weather condition, nor are they available everywhere. Singularly, the production of rammed earth and adobe requires inputs that can be made locally. Rammed earth is elaborated with moistened soil compacted in situ, temporarily shuttered, for creating large-sized blocks. On the other hand, adobe is a regular block composed of a plastic mixture of soil with a higher water content than rammed earth, and it is dried outdoors (Parracha et al. 2021). Earthen materials have been employed in construction for a long time ago. Currently, adobe and rammed earth are still used for new dwellings mainly because of the local availability of materials such as the mixture of soil with water and straw or grass as binders (Parisi et al. 2021). About a fifth of the population lives in adobe and rammed earth buildings; almost 50% are in developing countries (Seyfan 2010). Compared to current construction methods, these materials produce minimal waste and have a low risk of adding hazardous substances; also, adobe and rammed earth are easily recyclable. Therefore, its ecological footprint is minimal when the input materials are local since the energy used for its transport is also significantly reduced (Christoforou et al. 2016).

The selection of construction processes and materials with a high local availability favors the reduction of fuel consumption and lower emission of polluting substances. It is desirable to prioritize using natural materials that provide the population with comfort, energy savings, and preserving their living spaces (Benghida 2015). Adobe and rammed earth are affordable and, if properly maintained, are also long-lasting materials (up to 400 years) (Benghida 2015). These characteristics are significant in places where water scarcity and high temperatures require affordable alternatives for a growing population, as in many arid and semi-arid areas in the Global South.

### 6.1.3 Experiences with Adobe and Rammed Earth in Mexico

For thousands of years, adobe has been used to construct houses for the indigenous population in Mesoamerica (Gama-Castro et al. 2012) (Fig. 6.1). There are several examples in Mexico that evidence this long building process tradition: Huichapan, Hidalgo, in central Mexico (Gama-Castro et al. 2012), an archaeological zone in the state of Veracruz, in the country's southeast (Liberotti and Daneels 2012), in the state of Michoacán (Punzo-Díaz et al. 2015), as well as in another archaeological zone in Durango (Punzo-Díaz 2015). Also, there is corroboration of innovations in

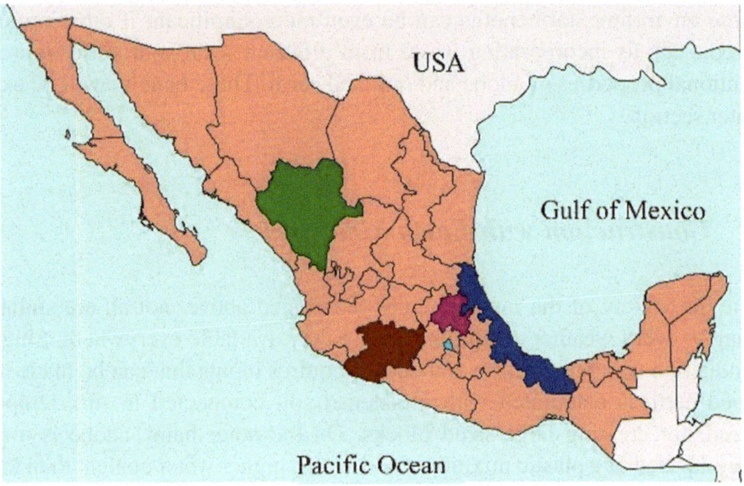


**Fig. 6.1** Some archaeological remains with adobe and rammed earth in Mexico. In green: Durango; brown: Michoacán; light blue: Mexico City; pink: Hidalgo; blue: Veracruz

the manufacture of adobe using nopal straw and corn as protective elements against inclement rain since the Aztec era (now the Mexico City area) (Martínez-Camacho et al. 2008).

Adobe houses are ubiquitous in rural Mexico, where their characteristics are modest and precarious, as a reflection of the economic power of its inhabitants. Due to their poor construction practices often have inadequate sunlight and ventilation conditions (Juárez-Sánchez et al. 2018). Fortunately, this eco-building process does not go unnoticed by contemporary researchers since its thermal benefits and natural lighting are welcome in urban constructions in Mexico's arid and semi-arid climates. In the city of Mexicali in northwestern Mexico, one hundred years ago, there were no air conditioning systems to mitigate temperatures of 42 °C, so this construction process was a passive technology for thermal regulation (Olvera et al. 2018). It is not a novelty in the construction industry to employ these materials to improve thermal conditions and durability in places as arid as the state of Sonora, where nopal mucilage was used, slowing down the degradation process by erosion (Martínez-Camacho et al. 2008). In northwestern Mexico, where temperatures can reach 54 °C, strategies were applied to improve livability conditions in homes for low-income people (Chan-López 2000). One of the materials analyzed to achieve this goal was adobe due to its low cost, little need for maintenance, and ability to reduce artificial cooling during the summer (Chan-López 2000).

The damage caused by hurricanes to adobe houses in the Guerrero state mountain areas (west-central Mexico) affected in the year 2013 a population with high marginalization (Hastings-García and Huerta-García 2015). In another location in the country's center, earthquakes occurred in 2017 affected many historic buildings built with adobe (Guerrero-Baca 2019). It was shown that the one- and two-story houses built with this material and that received maintenance were preserved intact; on the contrary, those constructions of adobe that employed reinforcements with industrialized materials had greater affectations (Guerrero-Baca 2019). From the preceding, it can be deduced that using these materials has many possibilities of being incorporated into the urban environment to meet the demand for affordable housing that satisfies comfort needs through passive technologies.

## 6.2 Properties of Adobe and Rammed Earth

Rammed earth and adobe have been building materials since the Neolithic period. The type of structures built with these materials ranges from simple dwellings to high-rise buildings. The durability of this type of earthen structure can be considerably long-lived under favorable conditions such as average annual temperatures above 18 °C and rainfall of around 250 and 650 mm (McHenry 1989).

### *6.2.1 Physical and Mechanical Properties*

#### 6.2.1.1 Water Absorption and Density

Earthen materials have a great capacity to absorb water, up to 3%, without exceeding this parameter (Arto et al. 2021). During manufacturing, the moisture content is an essential factor for the strength development of the material. To measure the water content in the mixture, a free fall test is performed, which consists of compacting the soil into a ball and letting it fall onto a flat surface from a height of approximately 1.5 m. When the ball of soil fractures into several segments, it is too dry, and if the soil remains in one piece, it is said that the humidity is adequate (Cravo 1989). Therefore, the density values of dry adobe range between 1500 to 2000 $kg/m^3$ (Loccarini et al. 2020). To guarantee the recommended humidity and compressive and tensile strengths more significant than 25 MPa and 0.25 MPa, respectively, the rammed earth structures' density values range from 1600 to 1900 $kg/m^3$ (Burroughs 2009). In general, an optimum humidity value allows compaction of the material, and the earth materials' dry density is an important parameter to obtain the strength required by the user (Steen 1971).

#### 6.2.1.2 Resistance to Compression and Tensile

These mechanical parameters allow materials such as rammed earth and adobe to be characterized even if they are brittle to obtain values that help to consolidate themselves as earthen materials. The compressive strength is obtained through axial tests performed perpendicularly to the direction of the layers of the soil samples. Due

**Table. 6.1** Compressive strength and tensile strength of clay

| | Mean compressive strength (Mpa) | Mean tensile strength (MPa) |
|---|---|---|
| Houses[a] | 1.32 | 0.17 |
| Land dividing walls[a] | 1.03 | 0.22 |
| All constructions[a] | 1.17 | 0.19 |

[a] Arto et al. (2021)

**Table. 6.2** Compressive strength and tensile strength of clay with construction additives in volume percentage

| Additives (%Wt) | Mean compressive strength (Mpa) | Mean tensile strength (MPa) |
|---|---|---|
| Cement[a] (8) | 3.2 | 0.45 |
| Lime[a] (4) | 3.3 | 0.49 |
| Puzzolanics[b] (10) | 2.5 | 0.25 |
| Coir fiber[c] (3) | 4.1 | 0.39 |
| Palm fiber[a] (0.75) | 3.3 | 0.45 |
| Barley fiber[a] (0.75) | 2.7 | 0.50 |

[a] Koutous and Hilali (2021); [b] Toufigh and Kianfar (2019); [c] Raavi and Tripura (2020)

to the manufacturing techniques of the adobe and rammed earth, the tests change, and the results obtained by the various studies can vary due to the absence of a standard procedure for determining this mechanical parameter (Raavi and Tripura 2020).

The tensile strength of the elements made with earth material is comparatively low due to cracking by their contraction. The reported values range from 0.15 to 0.25 MPa, making the relationship six times lower than the compressive strength tests (Table 6.1). Generally, soil material elements produce low flexural strength (Arto et al. 2021).

Different construction additives have been used for the stabilization of these materials and to obtain higher compressive strengths of the soil mixture (Table 6.2). They generally increase the resistance of 5–10 MPa with contents of 5–10% in volume substitution (Toufigh and Kianfar 2019).

The performance of earthen materials that can be observed in natural conditions varies. However, there are standardized parameters of mechanical characteristics for this type of earthen structure (Table 6.3).

### 6.2.2 *Geotechnical and Mineralogical Characteristics*

Different types of earthen materials are more durable than others, mainly due to the size of the particles and the proportion between them (Gomes et al. 2018). Generally, the particles of earthen structures are classified according to their size: gravel with

**Table. 6.3** Mechanical performance of earth materials

| Mechanical characteristics | Earthen materials | |
|---|---|---|
| | Not stabilized | Stabilized |
| Compressive strength[a] (MPa) | > 1 | 2–4 |
| Tensile strength[a] (Mpa) | 0.1–0.5 | > 0.2 |
| Traction strength[a] (Mpa) | 0.5–1 | 0.2–0.4 |
| Shear strength[b] (Mpa) | > 0.5 | - |

[a] Rigassi (1985); [b] Gomes et al. (2018)

a particle size greater than 2 mm, sand with a particle size of 0.02–2 mm, and clay with a particle size of 0.002–0.02 mm (Laborel et al. 2017). The durability of earthen structures is affected by excess gravel or clay; however, the proportion of clay in this type of material is fundamental since it intervenes in parameters such as water absorption, drying, and material density (Bell 1996). Adequate proportions of clay and sand bind between adobe or rammed earth structures since the sand particles are firmly embedded in the structural matrix when there is moisture (Ciancio et al. 2014).

Adobe and rammed earth with a suitable durability generally have a sand/clay ratio of 70–80% and 10–15%, respectively, while the rest is constituted by a minimum amount of gravel without construction additives (Valverde et al. 1997). On the other hand, the water used for the different earthen structures varies depending on the case; for adobe, the optimal amount is 20–30%, while for rammed earth, it is 10–15% (Arto et al. 2021).

#### 6.2.2.1 Composition of Earth Materials

Earth structures are mineral materials formed mainly by the disintegration of rocks combined with the interaction of various atmospheric agents. The main component of earth materials is silicate ($SiO_2$) and alumina ($Al_2O_3$). The chemical composition of this type of earthen structure reported in arid and semi-arid climates is presented in Table 6.4.

### *6.2.3 Thermal and Acoustic Properties*

Due to their excellent properties, adobe and rammed earth have been used in several parts of the world (Reddy 2022). Among the advantages of using earthen materials, noise, and thermal isolation can be highlighted, as well as fire resistance (Moscoso-Cordero 2010). Moreover, in semi-arid and arid climates, thermal characteristics in buildings are desired to control extreme temperatures, while acoustic isolation brings

**Table. 6.4** Chemical composition of earth materials in semi-arid and arid zones

| Semi-arid zones | | Arid zones | |
|---|---|---|---|
| Components | Wt.% | Components | Wt.% |
| $SiO_2$[a] | 66.13–79.36 | $SiO_2$[d] | 46.23–61.74 |
| $Al_2O_3$[b] | 11.85–14.38 | $Al_2O_3$[e] | 10.01–13.65 |
| $Fe_2O_3$[b] | 1.81–6.68 | $Fe_2O_3$[e] | 1.80–3.97 |
| MgO[a] | 0.15–0.34 | MgO[e] | 1.25–2.46 |
| CaO[a] | 0.41–0.73 | CaO[d] | 5.47–14.58 |
| CuO[a] | Trace | CuO[e] | Trace |
| MnO[a] | 0.15–0.33 | MnO[e] | Trace |
| $Na_2O$[c] | 0.24–0.46 | $Na_2O$[d] | 0.36–0.98 |
| $K_2O$[c] | 0.33–1.00 | $K_2O$[d] | 0.51–1.50 |
| $TiO_2$[a] | 1.09–1.27 | $TiO_2$[e] | 0.22–0.57 |

[a] Koutous and Hilali (2021); [b] Toufigh and Kianfar (2019); [c] Raavi and Tripura (2020); [d] Achiou et al. (2016); [e] Gomes et al. (2018)

comfort to housing. Although both construction processes have similar compositions, they differ in thermal and acoustic properties.

#### 6.2.3.1 Thermal Properties

One of the advantages of building with earthen materials in arid and semi-arid climates is that they remain dry through winter, which allows them to have good heat retention. In summer, they usually keep lower temperatures than outside, bringing freshness and comfort (Smith and Austin 1989). These characteristics are present in edifications using the traditional construction process. However, it was found that relatively recent modifications to the conventional method, for instance, using thinner walls, can reduce thermal comfort, especially in rammed earth houses where at least 45 cm thick walls are recommended (Portocarrero-Aguilar 2021).

Material thermal performance is commonly assessed in terms of R-values, a thermal resistance unit that indicates the material's resistance to transfer heat. Therefore, a higher R-value means better thermal insulation. Table 6.5 shows the thermal conductivity, specific heat, and R-value of the studied systems, along with other commonly used wall materials.

Adobe and rammed earth have R-values of 0.65 and 0.30 $m^2K/W$, respectively (Fuentes-García et al. 2016); however, these values vary depending on the mix's proportions, the soil type, and the aggregates. Nonetheless, both materials show better thermal insulation than concrete blocks and mass concrete, which present R-values of 0.22 and 0.13 $m^2K/W$, respectively (Fuentes-García et al. 2016). Different representative values are given by thermal conductivity, which measures the conduct heat capacity of materials, and the specific heat, which shows how much heat a material needs to increase its temperature. Focusing on R-values, both earth systems

**Table. 6.5** Wall construction systems' thermal properties

| Material | λ (W/mK) | $c_p$ (J/kg·K) | R ($m^2K/W$) |
|---|---|---|---|
| Adobe | 0.46 | 1480 | 0.65 |
| Concrete block | 0.91 | 1000 | 0.22 |
| Mass concrete | 2.3 | 1000 | 0.13 |
| Rammed earth | 1 | 600 | 0.30 |

Note: λ = thermal conductivity; $c_p$ = specific heat; R = R-value
Adapted from Fuentes-García et al. (2016)

are recommended for arid and semi-arid climates due to their thermal isolation, being adobe a noticeably better option.

#### 6.2.3.2 Acoustic Properties

The earthen structures also show excellent acoustic performance even without construction additives. Soil structure has a system of voids with a low percentage of connected pores, generating much less echo than other common walls made with different materials (Turco et al. 2021). Consequently, on average, earthen structures show a sound reduction index of approximately 50–65 dB for samples evaluated with thicknesses of 30–50 cm (Avila et al. 2021).

Better acoustic performance can be obtained by using construction additives in producing earthen materials; natural fibers, in general, are the most used to reduce noise because they show lower values of acoustic absorption coefficients than other construction additives (Chung and Blaser 1980).

## 6.3 Manufacturing and Construction Processes

Beyond the good properties of adobe and rammed earth, these construction processes have characteristics that keep them operational. This might be associated with different factors; first, the excellent accessibility of the raw materials, even in semi-arid and arid environments. Secondly, the equipment needed to make the bricks or walls is mainly wood hand-made molds. Finally, the process can be quickly learned and does not require skilled people (Costa et al. 2018). The two construction processes are discussed separately below.

### *6.3.1 Adobe*

A typical construction process consists of structural walls that pair adobe bricks agglutinated with mud as mortar (Fig. 6.2).

**Fig. 6.2** Adobe building structure

The bricks are manufactured in various dimensions, commonly 50 cm × 30 cm × 15 cm, and with varied granulometric compositions (Rivera-Torres 2012). They result in walls of different widths and heights. Most of these structures are built on stone foundations to prevent undermining of the walls (Austin 1984). Finally, the roof is commonly made of a wooden frame with eaves that protect walls from rain (Tanaçan 2008).

Focusing on adobe, bricks are chiefly made of clay and sand (Boudreau 1971). Also, straw is commonly used to prevent shrinkage and crack (Austin 1984). Other natural fibers are used to substitute this last material; for instance, in Mexico, dry grass is sometimes used (Matus et al. 2013). To create the brick molds, wood is mainly employed. Lastly, water is also needed during construction to make the soil mix.

To create the bricks, soil with enough clay and silt is needed to create a matrix where the sand particles can be solidly embedded (Brown and Clifton 1978). Therefore, the soil with the desired characteristics is searched close to the construction area. Later, the vegetal layer, of about 15 cm, is removed, and the soil is mixed with water with a shovel until it gets a proper consistency. Then, the natural fiber is added, and the mix is poured into the wooden molds; the mix is tamped by hand, paying particular attention to corners and removing the excess. After a drying day, bricks must be removed from the molds carefully to keep their shape. Subsequently, bricks are dried for three more days and turned on the side for uniform drying. Next, bricks are sun-cured for three to four weeks (Smith and Austin 1989), ready to be used or stacked. Finally, the wall is constructed like masonry, using adobe bricks and the soil mix as mortar.

### *6.3.2 Rammed Earth*

Rammed earth and adobe buildings have a similar construction process. Like adobe, rammed earth is chiefly used for structure walls. They are built upon local stone foundations to protect walls from the water and are usually covered with wood structures. In addition, the exterior walls are sometimes insulated with stucco. Therefore, the wall's construction process is the main object of interest.

Only proper soil is needed to build rammed earth walls, along with water and forms used in the construction process. An adequate soil has a good mix of different-sized particles, creating a dense matrix after compaction. Moreover, the soil must contain clay which acts as a binder to keep the other particles together (Easton 2007). Once adequate soil is obtained, the mixing process is essential to homogenize it. Mixing also provides a uniform humidity distribution inside the soil matrix (Maniatidis and Walker 2003).

To create the rammed earth walls, the selected soil is tamped in layers using wooden forms to shape it (Smith and Austin 1989). Wooden structures can vary in dimensions; the wall thickness depends on the wall height; the higher the wall, the thicker it should be. First, forms are set over the foundation, particularly their firmness due to the forces applied while tamping. Then the moist soil is poured over 10–20 cm layers and compacted using a tamper (Vicente-Val 2010). The process is repeated until the form height is reached and then dismounted (see Fig. 6.3). Later, the form is mounted above or alongside the finished section, and the process is repeated until the wall is complete.

In both construction processes, doors and windows can be created using lintels. Lintels can be shaped by collocating steel or wood bars directly over the opening, as can be appreciated in Fig. 6.3. Additionally, a correct assembling length to avoid bearing issues is required (Maniatidis and Walker 2003). The following section will discuss the sustainability of these construction methods.

**Fig. 6.3** Rammed earth construction process

## 6.4 Sustainable Practices and Current Regulations

### *6.4.1 Environmental Benefits*

Earth-based materials have potentially low environmental impacts (Fernandes et al. 2019). It has been previously mentioned that traditional materials such as adobe and rammed earth positively impact the natural environment when used as eco-building techniques. This impact is possible because the design and construction procedures use locally available resources based on their indigenous local knowledge (Chandel et al. 2016). Its application is also relevant in areas susceptible to seismic and hydrometeorological risks and where less favored people require affordable construction.

Through life cycle analysis studies, authors have found that in terms of embodied energy and carbon per kg of adobe produced, transportation needs and consumption of natural resources diminish, significantly affecting the system's environmental footprint of production (Christoforou et al. 2016). This environmental benefit can be improved if locally produced waste is used (Christoforou et al. 2016). Another study determined that using this type of low-energy consumption materials in the home helps mitigate $CO_2$ in the environment by up to 101 tons/year (Shukla et al. 2009). There are also reports of a reduction in potential environmental impacts by approximately 50% compared to conventional materials (Fernandes et al. 2019).

On the other hand, using materials such as rammed earth in homes represents implicit energy savings. Another of their environmental benefits is the short time it takes to recover the energy used, which is close to only 1.54 years; this is equivalent to a decrease of 101 tons/year of $CO_2$ (Shukla et al. 2009). It is relevant to consider that searching for other materials incorporated as additives or substitutes reduces the consumption of highly $CO_2$-generating inputs, such as cement (Arrigoni et al. 2017). At the same time, its incorporation into local materials also reduces the costs of transporting construction supplies (Chandel et al. 2016). One of the advantages of construction with adobe and rammed earth in arid and semi-arid areas is that it is unnecessary to use cement as an additive since weathering is not as harmful as it occurs in areas with cold climates (Sameh 2014).

Among their advantages, they can provide sustainable alternatives for environmental degradation, overexploitation of natural resources, deforestation, greenhouse gas emissions, and high energy consumption (Chandel et al. 2016).

### *6.4.2 Waste Employed in Sustainable Construction*

Using eco-friendly construction materials, such as adobe and rammed earth, can contribute significantly to achieving sustainable development goals, specifically targets 11.6 and 12.2 (UN 2022). Using such materials can reduce the consumption of raw materials and minimize waste and by-products (Losini et al. 2021), thereby

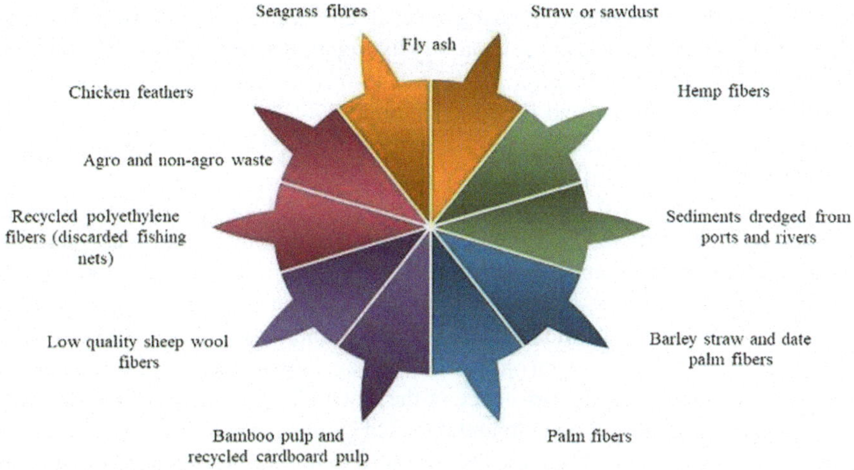


**Fig. 6.4** Variety of residues used in adobe and rammed earth

promoting a more circular economy and reducing the environmental impact of the construction industry. Additionally, they can improve indoor air quality and energy efficiency by regulating heat and humidity levels (Losini et al. 2021; Eslami et al. 2022). A bibliographic search was made of the main types of waste used to improve these materials' physical and environmental properties (Fig. 6.4).

Materials such as adobe and rammed earth are desirable in current construction because they reduce the consumption of raw materials and the cost of eliminating waste and by-products (Losini et al. 2021). They can also favor sustainability due to their ability to store heat and balance the humidity of the air inside buildings (Losini et al. 2021; Eslami et al. 2022). A bibliographic search was made of the main types of waste used to improve these materials' physical and environmental properties (Fig. 6.4).

The addition of these residues during the elaboration of adobe and rammed earth seeks to improve both their physical properties and the sustainable practices of this construction process and, indirectly, of other industries that seek to mitigate their negative impact on the environment. For example, the poultry industry produces a large volume of chicken feathers, mainly composed of keratin, a type of biopolymer from which fibers are obtained. Araya-Letelier et al. (2020) used these fibers in adobes. Although they did not improve their mechanical properties, they positively impacted their durability and resistance to damage by cracking and water erosion. Other agro-industrial residues, such as barley straw and date palm fibers, were used as stabilizers of rammed earth since they have a high percentage of water absorption (Koutous and Hilali 2021). As a result, the authors obtained a resistance similar to that of the reinforced rammed earth. However, it has environmental advantages since, by substituting cement or lime, there are not only economic savings but also environmental ones because the amount of incorporated energy required is reduced to manufacture these raw materials (Arrigoni et al. 2017). In another study, low-quality sheep's wool fibers were described as a very hydrophilic material. With its addition to the adobe mixture, it was possible to preserve its mechanical properties under humid environmental conditions and improve its durability (Parlato et al. 2022). Another example is sawdust in adobe, which improved its mechanical properties such as resistance (Costi-de Castrillo et al. 2021).

There are examples of other types of locally produced waste to improve this type of material's physical and environmental properties. Tensile strength has increased by using dredged sediments from rivers and ports; this is an alternative solution to valorize this type of waste (Hussain et al. 2022). Other residues used to optimize the properties of adobe include recycled polyethylene fibers (obtained from fishing nets). Bertelsen et al. (2021) not only achieved recycling of this type of waste but also improved the toughness and resistance of the material, significantly reducing adobe shrinkage cracking. Bamboo pulp and recycled cardboard pulp have been used due to their small diameter and because they are hydrophilic surfaces that create a porous network that reduces cracking caused by moisture loss. Adding cardboard pulp also improved adobe's flexural strength, toughness, and thermal insulation (Stanislas et al. 2021). Other fibers widely used as natural reinforcement in adobe bricks are obtained from the palm. Its benefits are increased mechanical properties and durability of these traditional construction processes (Eslami et al. 2022).

The above data is a good starting point to explore options for incorporating waste in materials such as adobe and rammed earth. These inputs can increase their ecological benefits and properties, such as resistance, hygrothermal capacity, or cost.

### 6.4.3 Some Challenges to Overcome

Adobe and rammed earth are associated with the need for intensive maintenance, preventing greater acceptance by the people (Kulshreshtha et al. 2020). Another disadvantage is when there are exposed to natural disasters such as earthquakes or hurricanes. Since earthen buildings are vulnerable, they must have reinforcement systems to avoid severe structural damage (Sameh 2014). Some studies suggest investigating the improvement of thermal comfort in facilities with these eco-practices since, when comparing their efficiency in winter, no considerable differences were found in comparison with thermal regulation in concrete structures (Zhang et al. 2016). On the other hand, although Mexican standards regulate the manufacturing and construction process with regular materials, it is worth considering some aspects not included until now (Table 6.6).

Table 6.6 summarizes some specifications employed in different parts of the world that could improve national and local regulations. From this, there is much to be done in Mexico to adopt sustainable construction methods and techniques. Some national standards define the context of sustainable structures and regulate their life cycle analysis. These do not include low-energy consumption materials (although they

**Table 6.6** Comparative among international and Mexican standards

| Material type | Specifications | Mexican standards/Limitations |
|---|---|---|
| Mud brick[a] | Soil type[a] | Sustainable buildings definition[g] |
| Rammed earth[a,b,c,d,e] | Particle diameters[f] | Life Cycle Assessment[g] |
| Press earth block[a,h] | Stabilization techniques[a] | Classifies materials not recommended[g] / Materials recommended are not included[g] |
| Adobe[b,c,d,e,f] | Design for earthquakes[a,f] | Factors that improve internal comfort[g] / Not consider what materials help with this improvement[g] |
| Poured earth[b] | Maintenance[a] | Specifications and test methods (earth compressed blocks)[h]/Other earth materials are not considered[h] |
| Pressed brick[b] | Protective coatings[a] | Conductivity methods[i] |
| Cob[c] | Termite protection[a] | Thermal resistance methods[i] |
| Cast earth[c] | Structural design[b,e] | Water absorption[i] |
| | Durability design[b] | Materials considered thermal insulants[i] |
| | Manufacturing process[c] | |
| | Operational performance (product installed)[c] | |
| | Design criteria[c] | |
| | Compressive strength[d] | |
| | Modulus of rupture[d] | |
| | Types of wall insulation[d] | |
| | Weathering protective measures[f] | |
| | Materials not recommended[g] | |

[a]Walker and Standards Australia (2002); [b]NZS 4297:2020; [c]ASTM (2016); [d]SRCA (2016); [e]NCh3332:2013; [f]Norma Técnica de Edificación e.080 adobe; [g]DOF (2013); [h]DOF (2016); [i]DOF (2011)

recommend which ones not to use), nor do they emphasize the need to use local materials to mitigate the environmental impacts of the construction sector.

Eco-building techniques with adobe for seismic zones are also highlighted in international standards. This could be good to consider due to the constant seismic structural damages reported in Mexican territory. Classifying and selecting materials (soil type, granulometry) and physical and mechanical tests are also necessary to improve the self-construction practices with these earthen processes. Mexican standards might also consider measures for the conservation and maintenance of earthen buildings, which are needed, especially in facilities established in vulnerable areas. From the above, it is recommended to incorporate standardized procedures to adopt the necessary tests that define construction techniques such as adobe and rammed earth in Mexico.

## 6.5 Conclusions

One of the prominent stigmas of the construction industry is its high capacity to emit pollutants into the atmosphere and its voracious consumption of natural resources. This scenario worsens with the growth of urban areas that demand more dwellings.

Furthermore, given the current economic conditions, people in countries of the Global South require affordable options that provide them with a livable and comfortable space. However, this need is not against sustainable alternatives that utilize local resources such as adobe and rammed earth. Their main advantages are that if they are built with reinforcing fibers (natural or from waste) and a suitable and standardized construction procedure, it is possible to reduce $CO_2$ emissions into the atmosphere, regulate the temperature inside the buildings, and increase their tolerance to humidity and rain. Another advantage is that these eco-building processes can obtain good mechanical properties without employing additives such as cement and its derivatives. This factor is especially relevant in places with arid and semi-arid climates, where it is possible to increase its durability against weather effects and use these eco-practices with a passive technology approach to improving thermal comfort.

Mexico is an example in the Global South where there is great potential for utilizing adobe and rammed earth as an alternative for sustainable construction in its urban spaces. As shown by the archaeological remains investigated in different parts of the national territory, culturally, the indigenous local people have used these materials to mitigate sudden temperature changes and take advantage of natural lighting and ventilation. In addition to their durability, when there is an adequate construction process, these buildings can better withstand the onslaught of natural disasters that are so common in the country, such as hurricanes and earthquakes.

Among its areas of opportunity is improving the techniques for making these eco-practices with the appropriate proportions of natural or synthetic fibers. This improvement can mitigate potential structural damage in seismic zones and reduce maintenance needs. The reuse of waste is desirable because it increases the sustainability of this self-constructive process and its durability, further reducing its negative environmental impact. Additionally, implementing national standards for its design and construction is highly recommended.

## References

Achiou B, El Omari H, Bennazha J, Albizane A, Daoudi L, Saadi L, Ouammou M, Alami S, El Maadi A, Chehbouni M (2016) Physicochemical and mineralogical characterizations of clays from Fez region (basin of Saiss, Morocco) in the perspective of industrial use. J Mater Environ Sci 7(5):1474–1484

Araya-Letelier G, Gonzalez-Calderon H, Kunze S, Burbano-Garcia C, Reidel U, Sandoval C, Bas F (2020) Waste-based natural fiber reinforcement of adobe mixtures: physical, mechanical, damage and durability performance assessment. J Clean Prod 273:122806

Arrigoni A, Beckett C, Ciancio D, Dotelli G (2017) Life cycle analysis of environmental impact vs. durability of stabilised rammed earth. Constr Build Mater 142:128–136
Arto I, Gallego R, Cifuentes H, Puertas E, Gutiérrez ML (2021) Fracture behavior of rammed earth in historic buildings. Constr Build Mater 289:123–167
ASTM (2016) Standard guide for design of earthen wall building systems, ASTM E2392/E2392M-10
Austin GS (1984) Adobe as a building material. New Mexico Geol 6:69–71
Avila F, Puertas E, Gallego R (2021) Characterization of the mechanical and physical properties of unstabilized rammed earth: a review. Constr Build Mater 325:121–135
Bell FG (1996) Lime stabilization of clay minerals and soils. Eng Geol 42:223–237
Benghida D (2015) Adobe bricks: the best eco-friendly building material. Adv Mat Res 1105:386–390
Bertelsen IMG, Belmonte LJ, Fischer G, Ottosen LM (2021) Influence of synthetic waste fibres on drying shrinkage cracking and mechanical properties of adobe materials. Constr Build Mater 286:122738
Boudreau EH (1971) Making the adobe brick. Fifth Street Press, Berkeley
Brown PW, Clifton JR (1978) Adobe I: the properties of adobe. Stud Conserv 23(4):139–146
Burroughs S (2009) Relationships between the density and strength of rammed earth. Proc Inst Civ Eng: Constr Mater 162(3):113–120
Chandel SS, Sharma V, Marwah BM (2016) Review of energy efficient features in vernacular architecture for improving indoor thermal comfort conditions. Renew Sust Energ Rev 65:459–477
Chan-López D (2000) Low energy housing, an architectural design and thermal simulation for extreme hot arid zones of Mexico: Mexicali B.C. In: Sayigh AAM (ed) World renewable energy congress VI, Chapter 138:686–689. Pergamon
Christoforou E, Kylili A, Fokaides PA, Ioannou I (2016) Cradle to site life cycle assessment (LCA) of adobe bricks. J Clean Prod 112(1):443–452
Chung JY, Blaser DA (1980) Transfer function method of measuring in-duct acoustic properties. I. Theory. J Acoust Soc Am 68(3):907–913
Ciancio D, Beckett CTS, Carraro JAH (2014) Optimum lime content identification for lime-stabilised rammed earth. Constr Build Mater 53:59–65
Costa C, Cerqueira Â, Rocha F, Velosa A (2018) The sustainability of adobe construction: past to future. Int J Archit Herit 13(5):639–647
Costi-de Castrillo M, Ioannou I, Philokyprou M (2021) Reproduction of traditional adobes using varying percentage contents of straw and sawdust. Constr Build Mater 294:123516
Cravo M (1989) Composition of a rammed earth sample from Silves Castle (in Portuguese). LNEC, Lisbon
DOF (2011) NOM-018-ENER-2011, Aislantes térmicos para edificaciones, características y métodos de prueba
DOF (2013) NMX-AA-164-SCFI-2013, Edificación Sustentable - Criterios y requerimientos ambientales mínimos
DOF (2016) NMX-C-508-ONNCCE-2015, Industria de la construcción -bloques de tierra comprimida estabilizados con cal -especificaciones y métodos de ensayo
Easton D (2007) The rammed earth house. Chelsea Green Publishing
Eslami A, Mohammadi H, Banadaki HM (2022) Palm fiber as a natural reinforcement for improving the properties of traditional adobe bricks. Constr Build Mater 325:126808
Fernandes J, Peixoto M, Mateus R, Gervásio H (2019) Life cycle analysis of environmental impacts of earthen materials in the Portuguese context: rammed earth and compressed earth blocks. J Clean Prod 241:118286
Fuentes-García RM, Valverde-Palacios I, Valverde-Espinosa I, González-Peregrín C (2016) Propiedades térmicas y acústicas en muros y revestimientos de tierra proyectada. In: CONTART 2016. La Convención de la Edificación, Granada, April 2016. Colegio Oficial de Aparejadores

y Arquitectos Técnicos de Granada, Consejo General de la Arquitectura Técnica de España, pp 89–97

Gama-Castro JE, Cruz-y Cruz T, Pi-Puig T, Alcalá-Martínez R, Cabadas-Báez H, Gomes MI, Faria P, Gonçalves TD (2018) Earth-based mortars for repair and protection of rammed earth walls. Stabilization with mineral binders and fibers. J Clean Prod 172:2401–2414

Gomes MI, Faria P, Gonçalves TD (2018) Earth-based mortars for repair and protection of rammed earth walls. stabilization with mineral binders and fibers. J Clean Prod 172:2401–2414

Guerrero-Baca LF (2019) Seismic performance of traditional adobe houses located on the slopes of Popocatepetl volcano, Mexico. Rev Gremium 11:104–117

Hastings-García I, Huerta-García G (2015) Reconstrucción y mejoramiento de la vivienda de adobe en la montaña de Guerrero, México. 15º SIACOT—ECUADOR, Seminario Iberoamericano de Arquitectura y Construcción con Tierra, p 11

Hussain M, Levacher D, Leblanc N, Zmamou H, Djeran-Maigre I, Razakamanantsoa A, Saouti L (2022) Reuse of harbour and river dredged sediments in adobe bricks. Cleaner Materials 3:100046

Juárez-Sánchez JP, Ramírez-Valverde B, López-Fuentes M, Ortega-López G (2018) Transformation of Mexican rural housing facing migration. The case of a locality in Puebla, Mexico. Rev Col San Luis 8(16):203–228

Koutous A, Hilali E (2021) Reinforcing rammed earth with plant fibers: a case study. Case Stud Constr Mater 14:e00514

Kulshreshtha Y, Mota NJA, Jagadish KS, Bredenoord J, Vardon PJ, van Loosdrecht MCM, Jonkers HM (2020) The potential and current status of earthen material for low-cost housing in rural India. Constr Build Mater 247:118615

Laborel PA, Aubert JE, Magniont C, Maillard P, Poirier C (2017) Effect of plant aggregates on mechanical properties of earth bricks. J Mater Civ Eng 29(12):13

Liberotti G, Daneels A (2012) Adobe in monumental architecture: chemical–physical analysis, archeology and 3D reconstruction to determine building techniques in the La Joya (Mexico) and Arslantepe (Turkey) sites. Bol Soc Geol Mex 64(1):79–89

Loccarini F, Ranocchiai G, Rotunno T, Fagone M (2020) Experimental and numerical analyses of strengthened rammed earth masonry arches. Comput Struct 239:1–20

Losini AE, Grillet AC, Bellotto M, Woloszyn M, Dotelli G (2021) Natural additives and biopolymers for raw earth construction stabilization—a review. Constr Build Mater 304:124507

Maniatidis V, Walker P (2003) A review of rammed earth construction. Innovation Project "Developing Rammed Earth for UK Housing", Natural Building Technology Group, Department of Architecture & Civil Engineering, University of Bath, p 12

Martínez-Bravo MM, Martínez-del-Río J, Antolín-López R (2019) Trade-offs among urban sustainability, pollution and livability in European cities. J Clean Prod 224:651–660

Martínez-Camacho F, Vázquez-Negrete J, Lima E, Lara VH, Bosch P (2008) Texture of nopal treated adobe: restoring Nuestra Señora del Pilar mission. J Archaeol Sci 35(5):1125–1133

Matus RA, Tizapa SS, Quiroz PC (2013) Caracterización experimental de las propiedades mecánicas de la mampostería de adobe del sur de México. Ingeniería 17(3):167–177

McHenry PG (1989) Adobe and rammed earth buildings: design and construction. University of Arizona Press, Arizona

Moscoso-Cordero MS (2010) El Adobe, sus Características y el confort Térmico. Congreso Internacional online Filosofía de la Sustentabilidad de Vivienda Tradicional "Transformando comunidades hacia el desarrollo local", p 71–75

NCh3332:2013. Estructuras - Intervención de construcciones patrimoniales de tierra cruda—Requisitos del proyecto estructural

Norma Técnica de Edificación E.080 ADOBE (2000) Lima, Perú

NZS 4297:2020. Engineering design of earth buildings. Standards New Zealand

Olvera D, Mendoza L, Guerrero L (2018) Adobe vernacular heritage in Mexicali, Mexico. In: Mileto C, Vegas LMF., García-Soriano L, Cristini V (eds) Vernacular and Earthen Arquitecture: Conservation and Sustainability. CRC Press/Balkema

Oyawa WO (2018) Eco-materials for developing countries. Struct Eng Int 14(3):208–212
Pacheco-Torgal F, Jalali S (2012) Earth construction: lessons from the past for future eco-efficient construction. Constr Build Mater 29:512–519
Parisi F, Tarque N, Varum H, Vargas-Neumann J (2021) Adobe constructions in the world: a first overview. In: Varum H, Parisi F, Tarque N, Silveira D (eds) structural characterization and seismic retrofitting of adobe constructions. Experimental and Numerical Developments. Springer, pp 1–15
Parlato MCM, Cuomo M, Porto SMC (2022) Natural fibers reinforcement for earthen building components: mechanical performances of a low quality sheep wool ("Valle del Belice" sheep). Constr Build Mater 326:126855
Parracha JL, Lima J, Freire MT, Ferreira M, Faria P (2021) Vernacular earthen buildings from Leiria, Portugal—Architectural survey towards their conservation and retrofitting. J Build Eng 35:102115
Portocarrero-Aguilar FI (2021) Arquitectura de tierra para mejorar el confort térmico climático de una vivienda eco-sostenible, en Huancas—Chachapoyas 2021 (Bachelor's degree thesis). Universidad César Vallejo, Perú
Punzo-Díaz JL, Rangel-Estrada D, Ibarra-Ávila E, Zarco-Navarro J (2015a) Primeros datos sobre el uso de adobe y cal en época prehispánica en la región Michoacana del río Balsas Medio, México. In: Achig-Balarezo MC (coord.) Tierra, sociedad, comunidad: 15° Seminario Iberoamericano de Arquitectura y Construcción con Tierra, pp 270–283
Punzo-Díaz JL (2015b) Revisando la arquitectura en el valle de Chalchihuiteña Guadiana, Durango. El caso del sitio arqueológico de la Ferrería. Cuicuilco 23(67):283–303
Raavi SSD, Tripura DD (2020) Predicting and evaluating the engineering properties of unstabilized and cement stabilized fibre reinforced rammed earth blocks. Constr Build Mater 262:120845
Reddy BVV (2022) Earthen materials and earthen structures. In: Compressed earth block and rammed earth structures. Springer Transactions in Civil and Environmental Engineering. Springer, Singapore
Rigassi V (1985) Compressed earth blocks: Manual of production. CRAterre-EAG GATE, 1
Rivera-Torres JC (2012) El adobe y otros materiales de sistemas constructivos en tierra cruda: caracterización con fines estructurales. Apuntes: Rev Estud Patr Cult 25(2):64–181
Sameh SH (2014) Promoting earth architecture as a sustainable construction technique in Egypt. J Clean Prod 65:362–373
Seyfan G (2010) Community action for sustainable housing: building a low-carbon future. Energy Policy 38(12):7624–7633
Shukla A, Tiwari GN, Sodha MS (2009) Embodied energy analysis of adobe house. Renew Energy 34(3):755–761
Smith EW, Austin GS (1989) Adobe, pressed-earth, and rammed-earth industries in New Mexico. Socorro, NM, USA: New Mexico Bureau of Mines and Mineral Resources. Bulletin 127
SRCA (2016) Title 14 Housing and Construction. 2015 New Mexico Earthen Building Materials Code 14.7.4
Stanislas TT, Tendo JF, Teixeira RS, Ojo EB, Komadja G, Kadivar M, Savastano-Junior H (2021) Effect of cellulose pulp fibres on the physical, mechanical, and thermal performance of extruded earth-based materials. J Build Eng 39:102259
Steen C (1971) An archaeologist's summary of adobe. El Palacio 7(4):29–38
Tanaçan L (2008) Adobe construction: a case study in Turkey. Archit Sci Rev 51(4):349–359
Toufigh V, Kianfar E (2019) The effects of stabilizers on the thermal and the mechanical properties of rammed earth at various humidities and their environmental impacts. Constr Build Mater 200:616–629
Turco C, Junior ACP, Teixeira ER, Mateus R (2021) Optimisation of Compressed Earth Blocks (CEBs) using natural origin materials: a systematic literature review. Constr Build Mater 309:125–140
UN (2022) The sustainable development goals report 2022, United Nations. https://desapublications.un.org/publications/sustainable-development-goals-report-2022. Accessed 1 May 2023

Valverde I, Ontiveros E, Sebastián E (1997) El tapial de las murallas de Granada. Rev Edif 26:58–63
Vicente-Val B (2010) Construcción con tapiales. Herencia rica del pasado. Cercha: rev Aparej Arq Téc 106:74–77
Walker P, Standards Australia (2002) HB-195-Australian Earth building handbook
Zhang J, Xu W, Li A, Zheng K, Zhang J (2016) Study on improving thermal environment and energy conservation of quadrangle adobe dwelling. Energy Build 129:92–101

# Chapter 7
# Sustainable Technologies Applied to Asphalt Paving by Reusing Polymeric Waste

**Iván Erick Castañeda-Robles, Francisco Javier Olguín-Coca, and Citlali Gaona-Tiburcio**

**Abstract** Countries of the Global South such as Mexico need to modernize their transport infrastructure to be increasingly competitive internationally. A resilient infrastructure promotes sustainable industrialization and encourages innovation. Two key objectives were identified: To develop reliable, sustainable, resilient, and quality infrastructure, including regional and cross-border infrastructure, and to support economic development and human well-being. To modernize the infrastructure and reconvert the construction industry, the adoption of clean and environmentally friendly technologies and industrial processes is proposed. In arid and semi-arid zones, the intense and prolonged periods of drought and erratic rainfall, as well as strong temperature variations, alter the road durability and resistance, and their maintenance requirements. That is why the asphalt concrete used in these areas must consider durability, resistance to cracking, low road noise, and low environmental impact. An area of opportunity that stands out here is the recycling of materials as an option to reduce the consumption of virgin materials, which impacts costs and maintenance. In this work, the properties of asphalt concrete modified with recycled polymeric waste were reviewed. A compilation of information is included that contemplates asphalt and concrete evaluation strategies through conceptual frameworks of sustainability, environmental impact, and circular economy for new road networks.

I. E. Castañeda-Robles (✉) · F. J. Olguín-Coca
Engineering and Architecture Academic Area, Autonomous University of Hidalgo State, Carr. Pachuca-Tulancingo Km 4.5, Col. Carboneras, C.P. 42184 Mineral de La Reforma, Hidalgo, Mexico
e-mail: ivan_castaneda@uaeh.edu.mx

F. J. Olguín-Coca
e-mail: olguinc@uaeh.edu.mx

C. Gaona-Tiburcio
Faculty of Mechanical and Electrical Engineering, Autonomous University of Nuevo León, Pedro de Alba S/N, Niños Héroes, Ciudad Universitaria, C.P. 66451 San Nicolás de los Garza, N.L., Mexico
e-mail: citlalli.gaonatbr@uanl.edu.mx

 
L. Lizárraga-Mendiola et al. (eds.), *Sustainable Spaces in Arid and Semiarid Zones of Mexico*, The Latin American Studies Book Series,
https://doi.org/10.1007/978-3-031-53418-8_7

It is concluded that recycled polymeric materials are a sustainable alternative to asphalt.

**Keywords** Asphalt concrete · Recycling · Reusing · Plastic · Durability · Low maintenance

## 7.1 Introduction

The importance of infrastructure for development is also a fundamental part of the 2030 Agenda and the Sustainable Development Goals (SDGs), adopted in September 2015 by the United Nations General Assembly (UN 2022). Goal 9 refers to building resilient infrastructure, promoting inclusive and sustainable industrialization, and fostering innovation. In this regard, investments in infrastructure (transport, water, irrigation, energy, and information and communication technology) have been fully recognized as key to achieving sustainable development and empowering communities in many countries.

The road infrastructure of the arid and semi-arid zones of Mexico, which represent almost two-thirds of the territory, is relevant for the economic development of the country, mainly because they involve a large sector of the national population. There, great challenges such as scarcity of water resources, marked climatic variations, and the need to conserve ecosystem services, among many others, must be faced. Economic and social activities in these areas require balance and rationality in using their natural resources.

One of the proposed solutions for pavements in arid and semi-arid areas is sustainable asphalt, based on the reuse of materials. Asphalt is a critical alternative in pavement construction. The search for sustainable asphalt pavements requires effective decision planning and strategies related to technical feasibility and environmental solutions. For example, the production of pavements incorporating solid polymers has been proposed to increase the reusing and recycling of polymers such as rubber and to improve the physical and mechanical properties of asphalt. However, the problem that occurs in asphalt pavements is their low vitality to fatigue and persistent alteration; these problems are caused by the incorrect dosage of binders-aggregates, their interaction, the method of placement, combination, and compaction of asphalt joints, as well as the behavior of its materials to temperature changes (Figueroa-Infante and Fonseca-Santanilla 2015).

There are several ways to make roads more sustainable and provide solutions to reduce the environmental impact of discarded polymers (Arroyo et al. 2018). Polymers are materials adequate for modifying asphalts due to their physical and chemical properties and high compatibility characteristics. They can be elastomers that improve the behavior of asphalt and asphalt mixture before loading and unloading cycles or thermoplastic elastomers that at high temperatures have a thermoplastic behavior and at low temperatures an elastic behavior (Lozano et al. 2020). Adding polymers is the solution to the problem presented in asphalt combinations, giving rise to thermoplastic materials with excellent thermal and mechanical properties, good performance, good chemical resistance, and lower impact resistance. The use of polymers as modifiers has been scarce, perhaps because of their costs or the mixing conditions; in general, modified asphalts are used when the needs of specific properties are above economic aspects (Vila-Romano and Jaramillo-Briceno 2018).

One successful example of reused polymers is granulated rubber (GR). It has excellent physical properties and benefits, such as good crack resistance, higher durability, lower maintenance, and lower road noise (Cheng et al. 2019). The amount of rubber material incorporated into roads can be significant, without modification of their mechanical properties (Ciro et al. 2015). In the 1840s, the first copyright of an asphalt binder modified with natural rubber appeared in England; later, in the 1960s, Charles H. MacDonald found a way to incorporate rubber from recycled tires (GCR) into asphalt, and this mixture was named asphalt-rubber (Yung-Vargas et al. 2016). Modifying asphalt with tire rubber increases performance and mechanical properties compared to natural binder, as rubber provides better strength to soundness, increases service life, and is used as an alternative that reduces costs (Kumar and Rahul 2020). The performance of GCR-modified asphalt depends on the rubber structure, which is modified by curing temperature, mixing time, and shear on developing granulated rubber's chemical and morphological structure in asphalt (Yao et al. 2016). The application of GCR in the asphalt pavement industry has received considerable attention in European countries such as Spain, Portugal, Switzerland, and Italy, because this material provides good mechanical properties and functional performance of pavement, besides being a way to recycle this waste product (Khaled et al. 2020).

In general, polymer-modified asphalt mixtures exhibit remarkable resistance to thermal cracking, rutting, water damage, and fatigue (Moussa et al. 2021). Consequently, in this chapter, we present a review focusing on the developments of asphalt mixtures for pavements, using recycled polymers as aggregate substitutes, fillers or combinations. A comparison between these asphalt mixtures against conventional asphalt pavements with similar mechanical properties was performed. Finally, we reviewed the strategies for evaluating asphalt through the conceptual frameworks of sustainability for new road networks.

## 7.2 Emerging Asphalt Pavement Technologies

### *7.2.1 Use of Recycled Polymers in Asphalt Pavements*

The number of investigations, proposals, and developments regarding using waste materials, mainly polymeric, to generate sustainable asphalt mix with better properties has increased in recent years. Some of this work is described below and includes the use of materials such as granulated rubber (GR), ground tire rubber (GTR), waste polyethylene (PE), polyethylene terephthalate (PET), water-based epoxy resin (WER), styrene-butadiene rubber (SBR), polyurethane (PU), bone glue (BG), conductive ethylene propylene diene monomer rubber (EPDM), styrene butadiene styrene (SBS), styrene-butadiene modified asphalt (SBSMA), thermoplastic elastomers (TPE), low-density poly(ethylene), (LDPE), poly(propylene) (PP), high-density poly(ethylene) (HDPE), and acrylonitrile butadiene styrene (ABS).

According to the APM (2017) report, in 2020, more than 29 million tons of post-consumer plastic waste were collected in the 27 European Union countries after the UK left the EU. More than a third went to recycling plants, more than 23% went to landfills, and more than 40% went to energy recovery operations. On the other hand, construction is one of the largest end-use markets for plastics (20%), but mainly for coatings and insulation. Recycling has been augmented by adding different types of polymers to asphalt bitumen (Wagner and Schlummer 2020). Field tests have proven that plastic waste used as an additive after proper processing can improve the life of roads and address some environmental problems. Its use in road construction is not new; it has already been used as high-density polyethylene (HDPE) pipe mesh crossings built together with polyvinyl chloride (PVC) wiring or as recycled HDPE pipes to form mats. Plastic waste is generally turned into powder and mixed (3–4%) with the bitumen. The durability of the roads incorporating powdered plastic waste is generally higher than that of conventional roads (Trimbakwala 2017). Although polymers can provide advantages and enhance some specific properties of asphalt, they can also affect others, or their re-processing can be expensive or have adverse environmental effects (Alqahtani 2022).

Some advances have been reported in the last decade; for example, the Dow company says having placed more than 26 miles of polymer-modified asphalt (PMA) pavement in various regions of the world, including Indonesia, India, and Thailand (Dow Corporate 2019). Similarly, there are reports of road works in the Durham Region, UK, where since 2018, they recycled plastic waste, including single-use bags and tire debris, in all their road resurfacing projects, and now rubber granules have been introduced into the mix (Durham County Council n.d.). Africa's first green road to incorporate plastic waste has been completed in Kouga Township in the Eastern Cape. A 300 m strip was covered with granular plastic-infused bitumen, reducing the amount required for the asphalt mix. They used about 1.5 tons of plastic, the equivalent of 1.8 million single-use bags, in just one kilometer of the road (News24 2019). PlasticRoad and Orbia, in collaboration with the Secretariats of Mobility and Environment of Mexico City, inaugurated in 2021 a bike lane in the Chapultepec Forest, which was made from more than 997,9032 kg of plastic waste (Orbia 2021). In India, according to information from the Municipal Corporation of Pune, plastic waste has been used in constructing test patches on several roads (Biswas et al. 2020). However, in practice, the use of plastics is still in the experimental stage, with gray areas, pending work, and ongoing research regarding standards and specifications. Further development is related to the presentation (powder, crushed, flake, or granules) and size of the plastics, the type, addition, and performance, among other aspects.

Noor and Ur-Rehman (2022) recently published a review about the use of plastic waste as modifiers of bituminous mixtures for flexible pavements, addressing the plastic problem, describing the characteristics of flexible pavements and their need for modification, the types of improvements, and the processing techniques. Appiah et al. (2017) examined the effect of blending waste thermoplastic polymers (HDPE and PP) in conventional AC-20 grade bitumen in various plastic compositions. It was found that they lowered the penetration value, increased the softening point, and improved the binder's overall dynamics and absolute viscosities. Their results showed that plastic waste-modified bitumen is a promising alternative recycling method for plastic waste management and a non-traditional modified binder for road construction (Appiah et al. 2017).

Kakar et al. (2021) summarized some of the published studies about PE waste and highlighted the importance of mixing temperature, amount of additive, time, and combining method. They mixed two types of PE waste, in granules and fragments, to analyze the effect on the mechanical and thermal behavior of the mixture. The resistance to high-temperature wear and rutting could be significantly improved with the addition of PE. In addition, the modulus at low temperatures was similar. However, the results were negative in terms of storage stability of the PE mixtures at high temperatures (Kakar et al. 2021).

It is well known that most single-use plastic bottles are made of PET. Investigations have shown that it can increase the resistance to the formation of furrows and improve rigidity, fatigue resistance, and surface appearance, but it can also affect negatively. For example, PET affects durability, binding properties, viscosity, and softening point or caused phase separation (Liang et al. 2019). Anwar et al. (2021) studied PET and PE in asphalt materials. They found that the stability of the plastic-modified asphalt pavement increased after 30 days while still meeting standard criteria with a plastic waste content up to 20%.

On the other hand, Sheng et al. (2018) studied the influence of different PU contents on the morphology and composition of the elements of the evaporation residues of emulsified asphalt, the thermal stability, the phase microstructures, and the network structure. They concluded that the emulsified asphalt modified with PU showed good mechanical and thermal properties. Yang et al. (2020) prepared waterborne-epoxy–SBR composite modified asphalt emulsions (WESAE) with different amounts of WER and characterized their workability, storage stability, and residual properties. It was found that incorporating WER improves thermal stability, high-temperature behavior, rheological properties, and SBR-modified asphalt emulsion residue adhesion. However, excessive WER affects the uniformity of particle distribution, storage stability, and workability of the recycled WESAE binder. Yao et al. (2022) indicated that emulsified asphalt's softening point, penetration, and ductility were improved with SBR, while flexibility decreased with WER. Luo et al. (2019) analyzed the effect of waterborne epoxy-modified emulsified asphalt (WEMEA), SBS, and a fiber stabilizer on modified emulsified asphalt. They found that SBR has excellent film-forming performance and suitable viscosity. Li et al. (2012) used the number density and density distribution maps to study the influence of different SBS concentrations and block proportions on the modified asphalt phase structure, considering the polymeric characteristic of the SBS molecular coating. In addition, they analyzed the adsorption behavior of the Saturated, Aromatics, Resins, Asphaltenes (SARA) asphalt components and SBS. Their results give further insight into the modification mechanism of SBS in asphalt and provide a reference for properly preparing SBS-modified asphalt. Chen et al. (2021) studied the stability of the phase structure of SBSMA (SBS Modified Asphalt) mixed with various materials. The effect of environment, dosage, type of SBS, type of asphalt, and processing technology in the structure and phase separation was reviewed too. They stated that several factors, mainly the characteristics of the raw materials, influence the phase structure stability. Both organic and inorganic materials can be cross-linked and swollen with SBS by grafting or direct incorporation, which improves the compatibility between SBS and asphalt and stabilizes the phase structure of SBSMA.

Li et al. (2019) synthesized a water-based curing agent for epoxy resin with a solid self-emulsifying ability. They analyzed the feasibility of using WEB as a high-performance cold binder. They concluded that the WEB resin decreased the emulsion's moisture sensitivity, increased rutting resistance, and significantly improved its mechanical strength, adhesion performance, and wear-life fatigue. Consequently, they considered that WEB could be a high-performance binder for cold recycling and asphalt mix paving.

Wang et al. (2015) used TPE (based on ground tire rubber, GTR) and PE for asphalt modification. With this, the compatibility between GTR and PE was improved with the dynamic vulcanization process, and the TPE stabilization in the asphalt was enhanced. Audy et al. (2022) recently evaluated 31 samples of plastic waste from different recycling companies in Australia and New Zealand as bitumen/asphalt modifiers. They developed classification criteria to identify recycled plastics suitable for road use. Its work determined that LDPE, ABS, and PET should preferably be used as bitumen/asphalt modifiers. Among them, recycled LDPE proved to have the highest affinity for bitumen as a polymer modifier, while ABS had the highest affinity for asphalt as an aggregate replacement.

Wu and Montalvo (2021) reviewed the feasibility of plastic waste reuse in asphalt pavements. One addressed topic was using plastics as an aggregate substitute or as modifier in the binder. Conventionally, plastic waste can be used in asphalt pavement through a dry or wet method. The dry process (as an aggregate substitute) incorporates the plastic waste particles (mainly rigid plastics with high melting points) into hot aggregates before adding a binder. The hardness and rigidity of the plastic particles contribute to the integrity of asphalt mixtures, playing a role similar to that of fine aggregates. This was investigated by Dalhat et al. (2019), who determined that the asphalt mixture containing finer aggregates performs better in asphalt mixtures. Nanjegowda and Biligiri (2020) revised the technologies and developments related to the GR use in asphalt paving applications. They addressed the wet and dry methods, their limitations and environmental aspects, rheological properties, and storage stability. They pointed out that the binder rheology was affected by the mixing temperature, rate, and time, but the operating or processing conditions of GR blend via the wet process could not be optimized. Rodríguez-Fernández et al. (2020) evaluated the effect of aging and the recyclability potential of modified asphalt mixtures with GR through the dry method and mechanical and rheological tests. They established that the polymer coating negatively changes the resulting GR performance. According to Colbert and You (2012), the combination of asphalt and plastics can be uniform in the wet method, and the plastic can be adequately dispersed in the binder. A comparison between the dry and wet processes in the production of waste polymer-modified asphalt mixes was conducted, and it was found that all combinations were comparable in terms of volumetric, stiffness, and moisture resistance (Ranieri et al. 2017). Due to the use of soft asphalt, these authors concluded that the dry method led to an exciting performance compared to the wet one.

### 7.2.2 *Use of Rubber*

Around 2014, an asphalt modifier derived from animal waste, such as skin, bone, and meat, came into use and was named Bone Glue (BG). Lv et al. (2021) analyzed its combined use with granulated rubber (GR) and found that the softening point and ductility increased while penetration decreased. Also, the combination of BG and GR increased the creep rate, rutting factor, and recovery rate. In addition, it diminished the creep modulus and the non-recoverable creep compliance. They also signaled that mixing them into asphalt can improve its high and low-temperature characteristics so that BG improves the storage stability of GR-modified asphalt while BG/GR modifier disperses evenly and stably in the asphalt matrix.

Han et al. (2019) combined three functional components in an EPDM rubber composite material with good interface bonding. They analyzed the mechanical properties of carbon fiber-reinforced EPDM rubber compounds and the thermal behavior of the composite material in high- and low-temperature environments. Authors determined that EPDM played a protective role, leading to good temperature durability, and excellent mechanical stability and heat-producing ability. In a recent work, Wei et al. (2022) studied the effect of adding EPDM rubber composite material on pavement structure response. For this, they placed the composite material under the asphalt layer in the pavement structure. They demonstrated that under the static load of the vehicle, the addition of EPDM can reduce the risk of cracking the asphalt pavement.

Mazzotta et al. (2020) studied the rheo-mechanical effects of rubber in asphalt mixtures by adopting a dry-hybrid technology that allows using rubber powder as a filler. They demonstrated that the presence of rubber influences the properties of the mastic and that the stiffness and the thermosensitivity of asphalt mixtures are related to the rheo-mechanical behavior of the putty and mortar.

Some authors have worked on developing asphalt mixtures for pavements using polymers and polymer residues as aggregates, substitutes for aggregates, fillers, or combined. Figure 7.1 shows some investigations presenting favorable results for the use of polymers. Improving the mechanical properties of the mixtures, both in the hot asphalt mix and in the cold mix, reduces the carbon footprint by betting on the use of materials and waste that seek to improve pavements.

Table 7.1 presents the results of several works that improve the mechanical properties of asphalt mixtures by adding tire rubber. Some of the mechanical properties shown are the reduction of deformations and distortions in the pavements, more excellent resistance to traction, and increased rigidity with optimal percentages. The

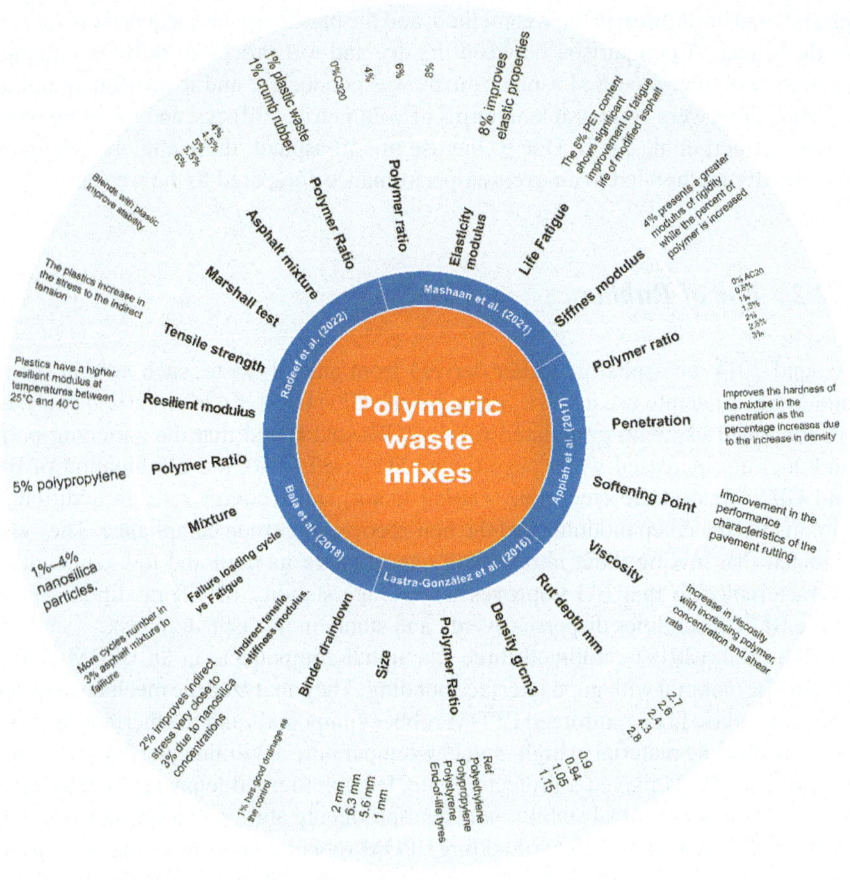


**Fig. 7.1** Characteristics of polymeric waste mixes used in asphalts

optimal rates evaluated are between 10 and 20%, improving the properties. Finally, the optimal range of 0.3–2.36 mm can be used when adding rubber to the asphalt mixtures.

## 7.3 Sustainable Asphalt Pavements

The European Road Federation-International Road Federation (ERF-IRF 2009) defines sustainable roads as roads effectively and efficiently planned, designed, constructed, modernized, and maintained through integrated environmental policies and retaining the expected socio-economic benefit in terms of mobility and safety. Sala et al. (2015) used the word “sustainable” in the context of asphalt pavement as presented in Fig. 7.2.

**Table 7.1** Improvement of asphalt mixtures by adding rubber

| Rubber ratio % | Particle size (mm) | Improvement | Author |
|---|---|---|---|
| 5 | 0.3 | Improved strength on asphalt | Hyun et al. (2019) |
| 2 | 0.6 | 8% helps improve high temperatures slightly; 10% improves high-temperature performance and ductility | Han et al. (2016) |
| 0.15 | 2.36 | Anti-deformation performance | Adday et al. (2021) |
| 5 and 10 | – | Greater strength in stiffness at low temperatures | Reza-Pouranian et al. (2020) |
| 8 and 12 | – | At 5%, these are negligible improvements. At 10% improved crack resistance | Bilema et al. (2021) |
| 3 and 6 | – | High stability 3%; low stability 6% | Unsiwilai and Santpetngam (2018) |
| 15 and 20 | – | Improves rutting | Liang et al. (2020) |

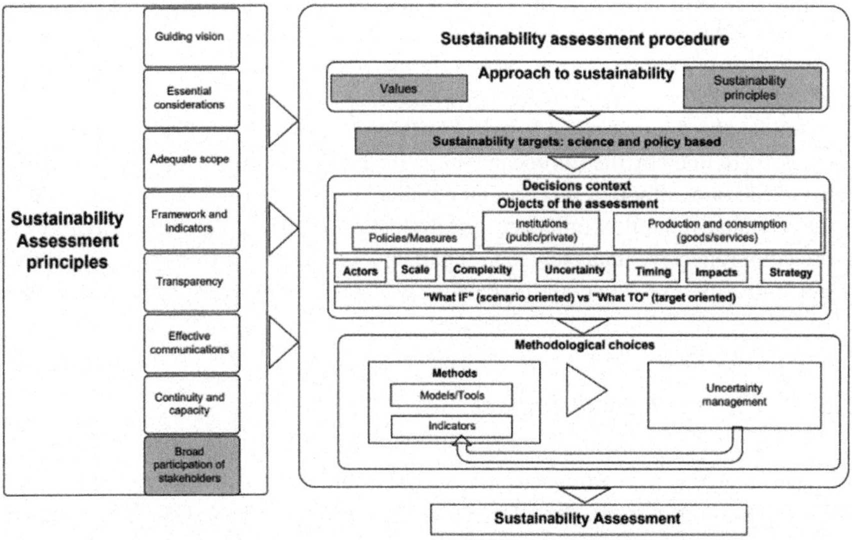


**Fig. 7.2** Schematic representation of the conceptual framework for sustainability assessment. Key elements, which distinguish sustainability assessment from integrated assessment, are reported in gray color. *Source* Sala et al. (2015)

The authors established the characteristics that pavement must have to meet the engineering challenges for which it was built, its preservation and conservation, and the use of economic, environmental, and human resources in terms of meeting the needs of comfort, safety, and mobility for people in their movements and their care for the environment.

Asphalt pavements contribute to the generation of greenhouse gases during their preparation, transport, and construction. In addition, there is evidence that the diversity of climatic phenomena, and their increased frequency and intensity, contribute significantly to the deterioration of asphalt pavements. So, it is essential to identify sustainability criteria, emphasizing the key areas that will contribute the most value to asphalt pavements and their compatibility with the environment, society, and the economy, allowing the generation of sectorial policies and action by decision-makers. The general framework within which a sustainable asphalt pavement should be evaluated may consider the following globally accepted aspects: the balance of the environment, society, and economy. Miller and Bahia (2010) proposed that the system sustainability assessment boundaries encompass five critical processes: an approach to sustainability, sustainability targets, objects of the assessment, methodological choices, and sustainability assessment (Fig. 7.3).

International practice on sustainability is very dispersed among the different components defined by both Green Roads and the European Union Road Federation. A work by Mendoza-Sánchez (2014) compared sustainability criteria applicable to highways in Mexico and identified the best practices developed on sustainable roads (Table 7.2). Current Mexican practice reaches only 33.3% of the requirements concerning the 57 concepts identified by Mendoza-Sánchez (2014). However, for the criteria defined by the European Union, the margin of compliance is more significant in Mexico, although it is only in number since there is no uniformity in all the criteria. Data availability influences the system boundaries: more data will allow for more comprehensive analyses. In comparison, less data will restrict the analysis to a single process in the overall life cycle.

### *7.3.1 Environmental Impact*

Climate change is a growing concern. Still, the number of vulnerable people in need of transportation by safe, durable, and sustainable means is more significant than ever. The potential socioeconomic benefits provided by highways include reliability under all weather conditions. These reduced transportation costs directly impact the fundamental right to mobility, which also allows free transit between regions, adequate distribution of food with accessible costs to the population, and access to essential services for health, safety, and other social services, such as emergency services, hospitals, clinics, schools, among others. But, in parallel, a high environmental impact has been associated with the construction of asphalt pavements, from clearing, leveling, or structure of the floor modifying the vegetation layer, which excludes other land uses and natural drainage patterns, to maintenance and use

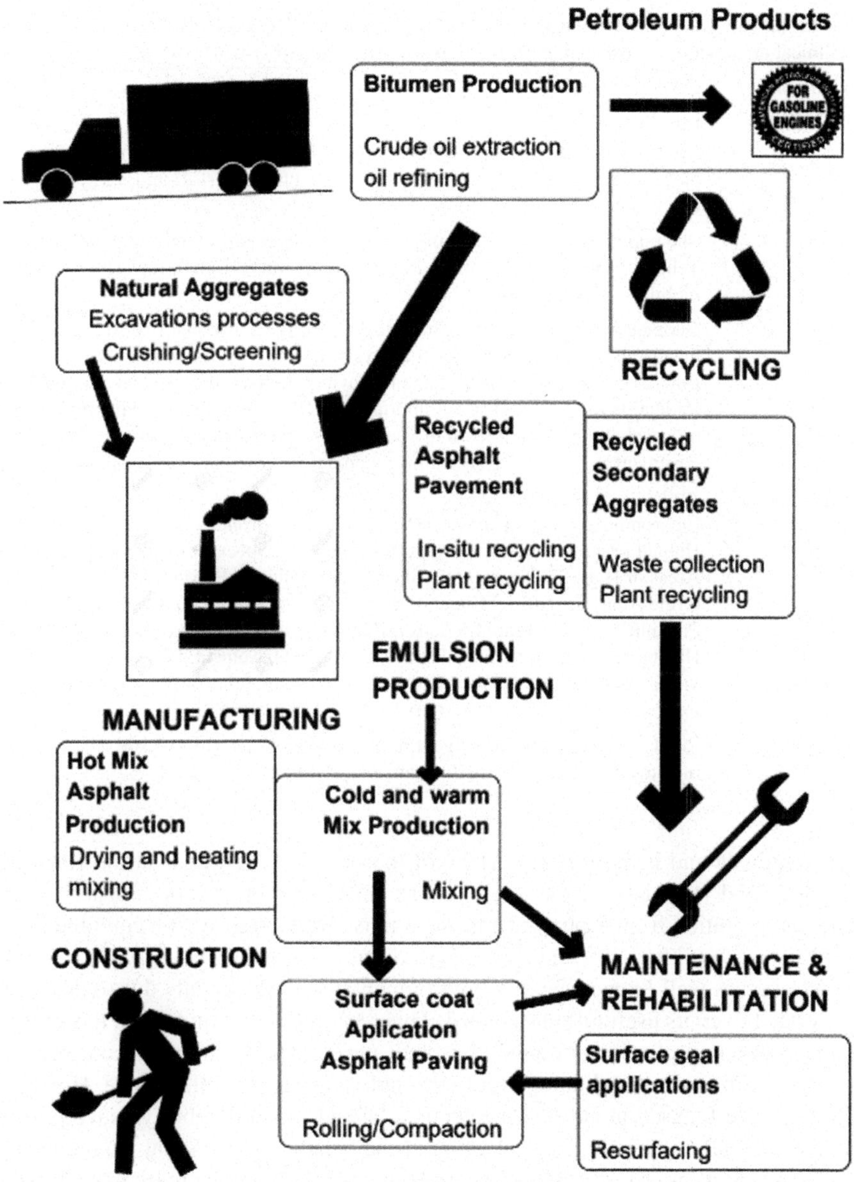


**Fig. 7.3** Typical system limits for asphalt pavement life cycle analysis. *Source* Miller and Bahia (2010)

**Table. 7.2** Sustainability criteria are identified in the best practices developed methodologies and technical documents on sustainable roads (Mendoza-Sánchez 2014)

| Components | Criteria |
|---|---|
| Economic | Life cycle cost analysis; Quality management system; Earthwork balancing; Contractor's warranty; Use of local materials; Long-lasting pavements; Best practices for road maintenance and infrastructure preservation; Energy efficiency |
| Social | Site maintenance plan; Planning in context; Quiet pavement; Light pollution; Vehicular emissions reduction; Pedestrian mobility; High occupancy vehicle mobility; Bicycle mobility |
| Technical | Geometric design; Life cycle inventory; Quality control plan; Waste management plan; Hydrological analysis; Use of recycled materials; Environmental management system; Environmental training; Recycling plan; Reduction of fossil fuel consumption; Reduction of emissions in paving; Warm mix asphalt (WMA); Record of water use in construction; Native vegetation; Pavement recycling; Pavement management system; Preservation of historical sites |
| Environmental | Environmental Impact Assessment; Strategic Environmental Assessment; Habitat Fragmentation/Ecological Connectivity; Special Considerations for Areas of High Environmental Value; Water Pollution Prevention Plan; Life Cycle Analysis; Noise Mitigation Plan; Surface Runoff Quality/Water Pollution Prevention; Surface runoff control; Reduction of fossil fuel consumption; Habitat restoration; Permeable pavement; Cold pavement; Landscape/Scenic views; Stormwater management; Alternate energy use; Carbon and nitrogen oxides sinks; Environmental training; Wildlife protection |
| Security | Road safety auditing; Intelligent transportation systems; Road surface maintenance; Pavement performance monitoring |

(Nanjegowda and Biligiri 2020). Many of these impacts can arise at the construction site and materials banks. Additionally, there may be adverse impacts on air and soil pollution from asphalt plants, dust, noise from construction equipment and explosives, fossil fuel use, and spillage (Rodríguez-Fernández et al. 2020).

Plastic has well-known negative environmental impacts, mainly due to the waste generated once its useful life has ended. These wastes are ubiquitous in the ecosystems. Recycling plastics in modified asphalt pavements for highway construction, conservation, and rehabilitation is a sustainable management alternative. However, studies have focused mainly on mechanical behavior and durability, leaving aside studies that allow quantitative evaluation of the benefits of recycling. To reduce the environmental impact of asphalt pavement construction and increase paving materials' sustainability, they must be evaluated through a series of ecological qualification tools such as life cycle analysis (Jamshidi and White 2020).

The life cycle analysis allows the environmental impact characterization of a road (Santero et al. 2011). Otherwise said, it is important to consider the costs associated with all phases of the life cycle of a road. The production of asphalt pavements derived from waste considers the extraction, acquisition, and production costs of raw and recycled waste materials. In some countries, standards and regulations have been developed to allow better use of materials and eco-inventions in recycled asphalt pavements, such as those proposed by the European Road Federation (ERF) and the Brussels Programme Centre of the International Road Federation (IRF 2009). These organisms establish that sustainable roads must consider the environment from the stages of planning to maintenance and rehabilitation work. For ERF-IRF (2009), one of the environmental objectives for sustainable roads is "to develop a methodology to audit the environmental quality of road projects during their entire life cycle, from planning to maintenance, including not only the planning of materials, construction, maintenance, service, and demolition but also evaluation of the optimization of the energy consumption of vehicles".

The transport sector collaborates to meet the objectives of the ERF-IRF (2009). The environmental analysis of the road works contributes to a sustainable solution, considering the pavements using recycled materials (Loprencipe and Cantisani 2013). To explore the effects of the construction companies and analyze the economic viability of investments to make the pavement product ecologically friendly, the concepts involved are environmental management in civil construction, technological protection, and sustainable performance. Environmental management is a global system that addresses the allocation of resources, the definition of responsibilities, and continuous evaluation of practices, processes, and procedures aimed at developing, achieving, implementing, critically analyzing, and maintaining the environmental policy established by the normative ISO 14001 (2015).

## 7.4 Conclusions

Asphalt pavements are structures that support transportation. Their construction, maintenance, and durability become strategies for a country's development and internal and external growth. The research presented herein provides the following conclusions.

The reports reviewed consider using recycled polymers as a material that improves the mechanical properties of asphalt for road construction, enhancing mechanical, chemical, and physical properties. So, knowing the different types of polymeric waste for proper use and application concerning the project's needs is essential.

The review of the information in this document seeks to promote elements to generate uniformity in all asphalt pavement development criteria with a definition of pavement sustainability and sets of regulations to be built. This needs to establish a system of tools, management controls, and quality assurance at all stages of the design and recycling of the waste with a view to sustainable asphalt pavements. The need to develop new asphalt pavements that innovate the use of recycled materials, environmental management systems, recycling plans, reduction of fossil fuel consumption and greenhouse gases emissions in paving, design of asphalt mixes with optimal mechanical and rheological properties, was also evidenced.

## References


Adday FA, Awad A, Alsaleh F (2021) Low-cost pavement by using solid waste, recycled aggregates, crumb rubber, and waste plastic for rural road. Int J Geomate 20:18–23

Alqahtani FK (2022) Technical assessment of Green lightweight concrete containing manufactured plastic aggregates. J Build Eng 50:104169

Anwar MK, Shah SAR, Alhazmi H (2021) Recycling and utilization of polymers for road construction projects: an application of the circular economy concept. Polymers 13(8):1330

APM (2017) Plastics—the facts 2017. An analysis of European plastics production, demand and waste data. Association of Plastics Manufacturers. https://plasticseurope.org/wp-content/uploads/2021/10/2017-Plastics-the-facts.pdf. Accessed 1 Sept 2022

Appiah JK, Berk-Boatenga VN, Tagbor TA (2017) Use of waste plastic materials for road construction in Ghana. Case Stud Constr Mater 6:1–7

Arroyo P, Herrera R, Salazar L, Giménez Z, Martínez J, Calahorra M (2018) Un nuevo enfoque para la integración de factores ambientales, sociales y económicos para evaluar mezclas asfálticas con y sin neumáticos de desecho. Rev Ing. Constr 33(3):301–314

Audy R, Enfrin M, Boom YJ, Giustozzi F (2022) Selection of recycled waste plastic for incorporation in sustainable asphalt pavements: a novel multi-criteria screening tool based on 31 sources of plastic. Sci Total Environ 829:154604

Bilema M, Aman M, Hassanc N, Halould M, Modibbo S (2021) Influence of crumb rubber size particles on moisture damage and strength of the hot mix asphalt. Mater Today: Proc 42:2387–2391

Biswas A, Goel A, Potnis S (2020) Performance comparison of waste plastic modified versus conventional bituminous roads in Pune city: a case study. Case Stud Constr Mater 13:e00411

Chen Z, Zhang H, Duan H, Shi C (2021) Improvement of thermal and optical responses of short-term aged thermochromic asphalt binder by warm-mix asphalt technology. J Clean Prod 279:123675

Cheng X, Liu Y, Ren W, Huang K (2019) Performance evaluation of asphalt rubber mixture with additives. Materials 12(8):1200

Ciro E, Parra J, Zapata M, Murillo EA (2015) Efecto de caucho reciclado en las propiedades de mezclas de caucho reciclado y polipropileno reciclado. Ing Cienc 11(22):173–188

Colbert BW, You Z (2012) Properties of modified asphalt binders blended with electronic waste powders. J Mater Civ Eng 24(10):1261–1267

Dalhat MA, Al-Abdul-Wahhab HI, Al-Adham K (2019) Recycled plastic waste asphalt concrete via mineral aggregate substitution and binder modification. J Mater Civ Eng 31(8):04019134

Dow Corporate (2019) Dow completes roads improved with recycled plastic. https://corporate.dow.com/en-us/news/press-releases/dow-completes-roads-improved-with-recycled-plastic.html. Accessed 12 Aug 2022

Durham County Council (n.d.) Road resurfacing programme. https://www.durham.gov.uk/article/18391/Road-resurfacing-programme. Accessed 21 June 2022

ERF-IRF (2009) The future of transport: strategic road infrastructure priorities–beyond 2010. European Road Federation and International Road Federation Brussels Programme Centre

Figueroa-Infante AS, Fonseca-Santanilla E (2015) Estudio de material reciclado para reparar fisuras y su aplicación en un pavimento en Bogotá. Épsilon 24:89–121

Han L, Zheng M, Wang C (2016) Current status and development of terminal blend tyre rubber modified asphalt. Constr Build Mater 128:399–409

Han S, Wei H, Han L, Li Q (2019) Durability and electrical conductivity of carbon fiber cloth/ethylene propylene diene monomer rubber composite for active deicing and snow melting. Polymers 11(12):2051

Hyun HK, Moon-Sup L, Soon-Jae L (2019) Performance evaluation of polymer modified asphalt (PMA) binders containing ground tire rubber (GTR). Int J Pavement Res Technol 12:215–222

IRF (2009) Innovative practices for greener roads. International Road Federation. https://es.calameo.com/read/000046992065d5df0cc24. Accessed 12 July 2022

ISO 14001 (2015) Environmental management systems-Requirements with guidance for use. International Standardization Organization
Jamshidi A, White G (2020) Evaluation of performance and challenges of use of waste materials in pavement construction: a critical review. Appl Sci 10(1):226
Kakar MR, Mikhailenko P, Piao Z, Bueno M, Poulikakos L (2021) Analysis of waste polyethylene (PE) and its by-products in asphalt binder. Constr Build Mater 280:122492
Khaled TT, Aboud GM, Al-Hamd RKhS (2020) Study the effect of adding crumb rubber on the performance of hot mix asphalt. IOP Conf Ser: Mater Sci Eng 737:012129
Kumar HV, Rahul BG (2020) Performance and cost analysis of modified bitumen binder for flexible pavement. IOP Conf Ser: Mater Sci Eng 912:062050
Li HP, Zhao H, Liao KJ, Li YG, Li XQ (2012) A study on the preparation and storage stability of modified emulsified asphalt. Pet Sci Technol 30(7):699–708
Li R, Leng Z, Zhang Y, Ma X (2019) Preparation and characterization of waterborne epoxy modified bitumen emulsion as a potential high-performance cold binder. J Clean Prod 235:1265–1275
Liang M, Xin X, Fan W, Wang H, Jiang H, Zhang J, Yao Z (2019) Phase behavior and hot storage characteristics of asphalt modified with various polyethylene: experimental and numerical characterizations. Constr Build Mater 203:608–620
Liang Y, Jones D, Harvey JT (2020) Laboratory evaluation of rubberized binder and mix containing a low content of devulcanized rubber modifier. Transp Res Rec 2674(11):53–63
Loprencipe G, Cantisani G (2013) Unified analysis of road pavement profiles for evaluation of surface characteristics. Mod Appl Sci 7(8):1–14
Lozano DA, Molina-Gómez F, Ruge JC, Moreno-Anselmi LA, Bastidas-Martínez JG (2020) Asphalts and modified dense asphalt mixtures with rubber of military boots. DYNA 87(212):120–128
Luo Y, Zhang K, Xie X, Yao X (2019) Performance evaluation and material optimization of Micro-surfacing based on cracking and rutting resistance. Constr Build Mater 206:193–200
Lv S, Tan L, Peng X, Hu L, Borges-Cabrera M (2021) Experimental investigation on the performance of bone glue and crumb rubber compound modified asphalt. Constr Build Mater 305:124734
Mazzotta F, Tataranni P, Simone A, Fornai D, Airey G, Sangiorgi C (2020) Multi-scale rheo-mechanical study of sma mixtures containing fine crumb rubber in a new dry-hybrid technology. Appl Sci 10(11):3887
Mendoza-Sánchez JF (2014) Criterios de Sustentabilidad para carreteras en México. Instituto Mexicano del Transporte, pp149. https://imt.mx/archivos/Publicaciones/PublicacionTecnica/pt392.pdf. Accessed 3 Sept 2022
Miller TD, Bahia HU (2010) Establishing a framework for analyzing asphalt pavement sustainability. Int J Pavement Res Technol 3(3):149–155
Moussa GS, Abdel-Raheem A, Abdel-Wahed T (2021) Effect of nanoclay particles on the performance of high-density polyethylene-modified asphalt concrete mixture. Polymers 13(3):434
Nanjegowda VH, Biligiri KP (2020) Recyclability of rubber in asphalt roadway systems: a review of applied research and advancement in technology. Resour Conserv Recycl 155:104655
News24 (2019) First 'plastic' road officially opened. https://www.news24.com/news24/community-newspaper/kouga-express/first-plastic-road-officially-opened-20191218. Accessed 9 Aug 2022
Noor A, Ur-Rehman MA (2022) A mini-review on the use of plastic waste as a modifier of the bituminous mix for flexible pavement. Clean Mater 4:100059
Orbia (2021) Plasticroad and orbia inaugurate the first plasticroad pilot in Latin America in Mexico City's Chapultepec forest. https://www.orbia.com/this-is-orbia/news-and-stories/plasticroad-in-mexico-city-en/. Accessed 25 July 2022
Ranieri M, Costa L, Oliveira JRM, Silva HMRD, Celauro C (2017) Asphalt surface mixtures with improved performance using waste polymers via dry and wet processes. J Mater Civ Eng 29(10):04017169–04017171
Reza-Pouranian M, Notani MA, Tabesh MT, Nazeri B, Shishehbor M (2020) Rheological and environmental characteristics of crumb rubber asphalt binders containing non-foaming warm mix asphalt additives. Constr Build Mater 238:117707

Rodríguez-Fernández I, Cavalli MC, Poulikakos L, Bueno M (2020) Recyclability of asphalt mixtures with crumb rubber incorporated by dry process: a laboratory investigation. Materials 13(12):2870

Sala S, Ciuffo B, Nijkamp P (2015) A systemic framework for sustainability assessment. Ecol Econ 119:314–325

Santero NJ, Masanet E, Horvath A (2011) Life-cycle assessment of pavements. Part I: Critical review. Resour Conserv Recycl 55(9–10):801–809

Sheng X, Wang M, Xu T, Chen J (2018) Preparation, properties and modification mechanism of polyurethane modified emulsified asphalt. Constr Build Mater 189:375–383

Trimbakwala A (2017) Plastic roads use of waste plastic in road construction. IJSRP 7(4):137–139

UN (2022) The sustainable development goals report. https://desapublications.un.org/publications/sustainable-development-goals-report-2022. Accessed 6 May 2023

Unsiwilai S, Sangpetngam B (2018) Influences of particle size and content on deformation resistance of crumb rubber modified asphalt using dry process mix. Eng J 22(3):181–193

Vila-Romano R, Jaramillo-Briceno JG (2018) Incidence of the use of polymers as modifiers of the asphalt. Rev Lasallista Investig 15(2):315–326

Wagner S, Schlummer M (2020) Legacy additives in a circular economy of plastics: current dilemma, policy analysis, and emerging countermeasures. Resour Conserv Recycl 158:104800

Wei H, Ma Z, He X, Han S, Jiang B (2022) Structural response analysis of conductive ethylene–propylene–diene monomer rubber composite pavement under validated temperature field. Constr Build Mater 328:127094

Wu S, Montalvo L (2021) Repurposing waste plastics into cleaner asphalt pavement materials: a critical literature review. J Cleaner Prod 280(Part2):124355

Yang J, Zhang Z, Fang Y, Luo Y (2020) Performance characterization of waterborne epoxy resin and styrene-butadiene rubber latex composite modified asphalt emulsion (WESAE). Coatings 10(4):352

Yao H, Zhou S, Wang S (2016) Structural evolution of recycled tire rubber in asphalt. J Appl Polym Sci 133:42954

Yao X, Tan L, Xu T (2022) Preparation, properties and compound modification mechanism of waterborne epoxy resin/styrene butadiene rubber latex modified emulsified asphalt. Constr Build Mater 318:126178

Yung-Vargas YW, Córdoba-Maquilón JE, Rondón-Quintana HA (2016) Evaluation of abrasión of a modified drainage mixture with rubber waste crushed (GCR). Tecnura 20(50):106–118

The manufacturer's authorised representative in the EU is Springer Nature Customer Service Centre GmbH, Europaplatz 3, 69115 Heidelberg, Germany. If you have any concerns regarding our products, please contact ProductSafety@springernature.com

Printed and bound by CPI Group (UK) Ltd, Croydon, CR0 4YY
01/10/2026
02225883-0004